The Ramblings of someone who has spent too long in the electronics business

Books by: Seth Pittham

The Ramblings of someone who has spent too long in the electronics business

The Ramblings of someone who has spent too long in the electronics business

Volume one

By:

Seth Pittham

Contents

Forward

What's this book about?

I was chatting to a friend, Andy Durrant , who I got to know as a result of his reading of one of these articles. He kindly wrote an introduction for this book. He sat himself down in my workshop in Spain, taking up valuable bloody bench repair space. Now, he is an interesting guy and done lots of interesting stuff, including writing a couple books. He started to explain that, in the modern techie era, publishing a book was a relatively straight forward thing to do. This book is his fault. I can give you his address if you want.

As part of my very limited advertising campaign ventures, I write short articles that are relevant to the electronic repairs that I do, for local magazines and papers. Coupled with a small advert at the bottom of each article, this seems to capture people's attention. I get customers coming in or ringing up saying "I saw your article". They never say "I saw your advert". As a result, I continue to write stuff. Andy suggested that it may be a good idea, before I depart this current existence (he may know more than I do), to collate the articles and publish a short book. So, there we have it. I could print off some copies and give them to friends, on the basis that they are all too tight to buy their own poxy bloody copy. The articles are very much varied and related to technologies old and new, engineers and people that I have admired over the years. I purposely don't go into great depth, mainly because there is a limited amount of space in some of the magazines. And, I try and keep things a little light hearted so that the non-technically minded individual can gain a little insight into the world of electronical bits and bobs.

As a kid, I grew up reading magazines like Practical Electronics, Wireless World but the best was Television. This was aimed at the industry's engineers, giving tech tips etc. One chap that used to write a few pages every month was Les Lawrey-Johns. He wrote slightly tongue-in cheek, was informative, brilliant. He enthused me into the repair business. I go into my background in a little depth at the back of the book, which you can read if you haven't already consigned this book to propping up your Juke Box. As child, being taken into local electrical repair shops by my Grandfather, I always thought how good it would be to have a shop like that, wandering around in a brown workshop coat grumbling at things and people. A little under 60 years later, what have I got?! Careful what you wish bloody well wish for. A business called Zeta Services, in Zurgena, South West Spain.

In these articles we will nip back and forward in time, covering a couple hundred years. All I can ask for is that you find some of this stuff interesting and informative. As I have said, you don't need to be a rocket scientist to read all this stuff. My aim is to provide a small insight into a world that, like it or not, determines our lives.

Introduction:

To get the very best from this volume, you need, or at least it would help to understand a little about the Author's quirky nature, which is reflected in his writing style.

Firstly the author has a wealth of electronics knowledge and his enthusiasm for the subject flows freely in his collected articles. Whether you're a like minded electronics wiz or only have a passing interest, you won't fail to be drawn in by the rich almost 'Wikipedia' knowledge within; the difference being the author can fill in the gaps where 'Wikipedia' stops.

Secondly, Seth has a quick dry sense of humour, this can easily be deduced by any amateur detective that visits his workshop; simply read the numerous posters and notes scattered around the place; warning that YOUR stupidity is going to cost you extra on your bill and that stupid questions are highly likely to receive equally stupid responses.

Humour aside Seth is one of the most lovable, sincere natured human being's you are ever likely to meet, to be known by him as a 'friend' is truly an honour.

In summary, the notice on the outside of his Emporiums door should read as follows.

"Enter at your own risk, think before you speak and those of a sensitive disposition are probably best not to enter at all."

Andy M Durrant February 2019.

Chapter one

Let's kick off with a quick review on recording, audio stuff and things.

Recording techniques over the years. Part 1

The start of an exciting 3 part series!

A customer recently asked me to transfer some 5" reels of ¼" tape to MP3 memory stick, a recording made 58 years ago of his daughter singing away, memories that we all have. I heaved my 1963 Ferrograph 633 tape machine onto the bench and connected it to a PC. Away we went. Two massively different technologies, happily relating to each other, slowly falling in love.

Humans have been obsessed with recording and saving the human voice since around 1877, when a chap called Edison visualized a cylinder, about 8" long and 3" diameter. He covered it in a thin layer of tin and knocked up a mechanism that would rotate the cylinder. The other part of the machine consisted of a funnel mounted on a sort of lathe screw. At the narrow end of the funnel was a needle which pressed against the tin coated cylinder. As the cylinder rotated, hand cranked, a gear slowly moved the funnel along its length. Edison spoke loudly into the funnel, which vibrated the needle, which in turn left a modulated (wobbled) spiral grove on the cylinder. When the funnel was returned to the beginning of the cylinder and the needle rested in the grove, as it rotated, the recorded words could be heard! The first ever recorded words were "Mary had a little lamb, she also had a duck....." Well sort of, I added the duck bit.

This was all good, but the cylinders were bulky, hard to copy and quality of sound was rubbish. Edison lost interest in this and went off to invent the light bulb. In order not to infringe on patents, in late 1899, Mr. Berliner developed a system that used a similar horn, funnel arrangement, but this time the recording was on a flat disc with the grove spiral running from outside to the centre. The metal disc was coated with a hard ink and when dipped in acid, the surface was eaten away, leaving a clean grove. With a few developments, the disc could be now used as a "Master" to make copies. It was placed in a steam press where it was squeezed onto a hard shellac compound, similar to the old 78s some of us remember!

Over the following 15-20 years, the technique was developed and standards such as the 78 RPM, 10" and 12" disc were agreed. Popularity grew in wanting to own a gramophone. Recordings of Opera and Dixieland jazz were being sort after. But, the whole system was still very mechanical, no electronics and microphones as yet. The big horn used to collect the sound to be recorded, dominated the recording studio.

Let's take trying to record a small ensemble. Different instruments produced different volumes of noise, so, the loud instruments such as drums and piano, would be placed furthest from the horn, woodwind players would be closer and the poor old violins would have their heads stuffed inside the horn, as high frequencies did not record well. Bass notes were also a problem. Double Basses didn't provide enough volume, so instruments like Tubas, bass and contrabass Saxophones were used to add a bit of grunt.

As electronics and valve technology stormed ahead from the mid 1920s, amplification, microphones made record recording much easier. The band, orchestra and singers, could have their own microphones and now a "balance" could be made. Recordings were still made by "cutting" a metal disc. This was now the master from which other originals and copies could be made. Development of the "cutting head" was at the forefront, both in the US and Britain. Columbia, Victor and Western Electric had their own designs, but it was a British chap called Mr. Blumlein (who I talk about later in this book), working with HMV and EMI, who came up with a high quality system, whose principles are still used today. So there.

When in the recording studio, "cutting" a disc was not for the feint hearted. The disc itself cost around 40 quid in the '50s, plus technician and producer's time. You set the band up, a quick run through of what was to be recorded. Then, on with the "Red Recording" light and it had to be done in one take. No tape editing or autotune! Any cock ups, duff notes, meant a wasted disc and studio time, adding up to a lot of money. If a musician made mistakes, he would be black listed or worse. I had the privilege of playing Sax in Ken Mackintosh's dance band and some of the stories as to what went on were legendary!

Next coming up, how Tape Recording changed things.

Recording techniques over the years – Part 2

In part 1, I wrote about the beginnings of sound recording, centered around discs. In this part, we will look at how tape recording developed.

Recording onto discs was expensive and the equipment cumbersome. In the late 1800s, an American developed a machine named the Poulsen Telegraphone. This used reels of thin steel magnetic wire. Using motors running at a set speed, the wire passed at some velocity, across the surface of a recording head. The head, basically an electromagnet, was supplied with an amplified signal, say from a microphone. This signal modulated the magnetic structure (domains) of the wire. When the wire was wound back and replayed across the head, the magnetic imprint of the wire caused a signal to appear at the output of the head. This small signal, when amplified, gave you a sort of authentic reproduction! Amongst other drawbacks, the steel wire rusted and was heavy.

In the '20s, a German chap, Kurt Stille, came up with a variation on the wire recorder, this time using a very thin metal tape. Tape was 6mm wide and about 0.08mm thick. He called it a Blattenerphone. Great name. These machines were not to be messed with. The reels of steel tape weighed a ton and the speed that the tape passed the recording heads was around 5 foot a second! When the tape broke, it was like being sprayed with razor blades, decapitating anyone in a 100 mile radius. There was an improvement in sound over the wire system. The BBC bought a few but realised it was a stop gap.

In the '40s, two companies, Ampex and 3M, in the U.S. set about taking tape recording to the next stage. A paper tape magnetically doped was designed. This was a big improvement over the steel tape. At the end of WW2, an engineer, John Mullin was sent to Germany to look at the advancements in recording made there. He came back with some AEG equipment, tapes and set about improving the systems. Bing Crosby who was a massive recording star also helped out as anything that would promote domestic reproduction (sound that is) would be huge business for record companies and so on. In '47 Bing Crosby recorded several shows and the tapes allowed broadcasting all over the U.S. He put his hand in his pocket and invested $50,000 in the Ampex Company. This started tape recording as we now know it.

The standard of ¼ inch wide tape was adopted. This was now made from thin plastic polymer, with a magnetic oxide coat on one side. In the '50s, professional recording machines such as the BRT2 developed by EMI, were running the tape at speeds of 15 or 30 inches a second, passed the recording and replay heads. The tape could be cut and edited and the quality was great. The Beatles at Abbey road studios recorded early material on these machines. Other manufacturers such as Philips, Leevers-Rich and Swiss made Studer made superb machines.

In the '50s, home recording took off. Tape recorders were coming down in price and you could afford a 5" reel of tape. The standard domestic machine ran at a tape speed of 3 ¾ inches a second. It gave you about 20 minutes a side of tape. The tape was a ¼ inch wide and the area of the recording head was a little under 1/8 of an inch. The head being placed at the edge of the tape, meant when you finished recording on one direction (track on the top of the tape), you could turn the reels of tape over and record on the lower track in the other direction. This doubled your recording time. BSR as well as making record players made great budget tape machines. The more expensive semi-professional machines had higher tape speeds which increased the quality of recordings. Stereo was also available and some machines had a 3rd replay head. This was fitted alongside the recording head, so you could listen to what was being recorded in real time, in order to check levels etc. The played back sound was delayed by a fraction of a second, the time it took for the tape to pass from recording head to play head. It gave you an echo effect! This facility was used throughout the recording industry. The famous WEM CopyCat Echo machine used a loop of tape and several replay heads to give a range of effects. Any self respecting Rock n Roll band had one!

In the next part....Cassettes! Bet you can't wait.

Recording techniques over the years. Pt. 3 The cassette!

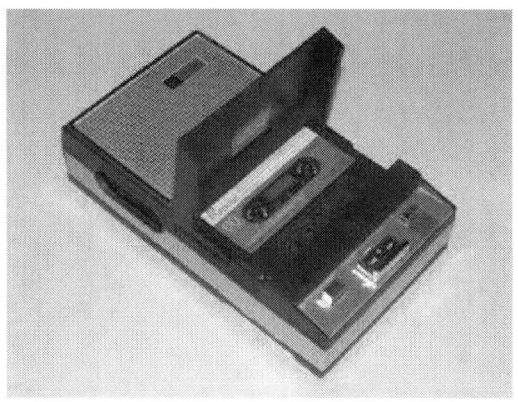

Part 2, I wrote about how recording sound on tape in the '50s-'60s had taken over from recording direct to metal discs and it's resulting popularity in the domestic market. The next breakthrough was the Compact Cassette.

RCA in the 'States recognised that lumbering big tape recorders and reels of tape around was not the most convenient of pastimes. They came up with a cased set of two small reels of tape, calling it a cassette. It was about 5" x 7", still too big and it flopped. In '63, Philips had developed and released the Compact Cassette as we know it. It was shown at the Berlin Radio show and released into Europe in '64. A massive hit, the Compact Cassette consisted of a plastic case, 4" x 2.5", containing 2 reels of 1/8" wide tape. The cassette had two holes which fitted onto the players' hubs for take up and winding back and forth and the tape was exposed at the front so that the recording / playback head, capstan and pinch roller could make contact with the tape head.

The tapes width was half that of the standard "open reel" type as was the speed that the tape travelled (1,7/8" per second). Although the design was subject to patents, Philips agreed to allow Sony to manufacture, which lead to Japan dominating the market. Philips released a great little Cassette machine in '64 called the Carry-Corder, type EL3302. It was an iconic design, simple to use with one control. Up for play, left and right for wind forward and back, this was a copy of the BSR tape decks popular in '50s.

Because the tape was narrow and moved slowly across the recording - playback head of the machine, audio quality, bandwidth and signal-to-noise was not great, but was ok to record Uncle Jack down the boozer on a Saturday night. Initially, the tapes magnetic composition was made of ferric oxide, fine rust if you like. Light brown in appearance and not too durable. Toward the late '60s companies such as TDK, 3M etc. were coming up with new coatings made from Cobalt and Chromium Dioxide. These gave a much better recording quality and lower background noise. Dolby Labs had designed a system whereby the audio signal was processed, minimizing the annoying hiss which could be heard during quiet passages. The cassette's quality now matched that of the LP record.

Because there were several types of tape composition, machines had to be adjusted to get the best out of the tape. Changes in "Bias" (method by which the sound was recorded onto the tape) had to be made for Standard, Chrome etc... Earlier machines had switches on the front panel so you could play around, whilst more expensive machines made use of the little cut-outs in the rear of the cassette housing which when placed in the cassette deck, married up with little switches, automatically setting the correct conditions. You also had different types of Dolby to switch, A, B, C and latterly D.

A big boost in sales came with Sony's release of the Walkman in early '80. This was a personal cassette player aimed at people who indulged in that stupid pastime of jogging. All sounded great until you started to jog however. The movement upset the rotation on the little capstan flywheel whose purpose was to move the tape at constant speed and the music wobbled all over the place. Later machines had two flywheels working in opposition, so the effects of your jogging cancelled out. Hands up who owned a shoulder mounted GhettoBlaster...admit it. The humble cassette allowed you strut your stuff down the road, listening to Motown. The weight of the machine was designed to balance out the chip you had on the other shoulder.

Companies such as Nakamichi and Revox made some excellent semi professional machines that managed to squeeze every last drop of quality out of the cassette. Sony & Teac developed a variation called the Elcaset. This was a much larger cassette with ¼" tape with a speed of 3-3/4" inches per second. Quality was great, but with the introduction of CDs in early '80s, the cassettes' days were numbered. So we say goodbye to the C30, C90 and C120.

Chapter Two

Analogue vs Digital   The Audio World
(bit of a follow on from the Recording Techniques 3 part chat)

A donation of a nice 1960 Ferguson Radiogram (pictured at the end of this chapter) got me thinking about the re-immerging of analogue recording techniques, vinyl records etc. Having got the set working to an "out the shop" condition, everyone who had walked in and heard it playing away in the background has commented "oh that sounds nice". Sure it does too. Not the most HiFi, but good on the ear, warm and comforting!

In the early '80s, CDs and digital recording started to take over from the old vinyl albums and studio tape mastering machines. All good stuff, but there has always been a hard-core school of thought that says the resulting sound is not as nice to listen to. To record and process the raw analogue musical signal to make a CD, it first has to be chopped up and made into electrical 1s and 0s, similar data format used by emails, phone calls etc. The analogue signal is incredibly complex and has to be chopped into very fine pieces in order for the complexity to be reproduced by your CD player and to sound half decent. The human ear and brain are very sensitive to tone and dynamic range (loud and soft). You have the ability to hear a quiet pin drop, all the way up to 5 Spaniards playing dominoes. There is a varying degree of loss of detail in the conversion of the signal into a digital format (quantisation loss). Purists say that it is noticeable. I agree. Try listening to an original recording of a piece of music, then the MP3 version. MP3 format has an even higher degree of loss of detail of all the audio formats.

In order to try and create this wide dynamic range when recording a band for example, so it sounds like the real thing on playback, the audio needs to be processed. Recording systems and your own HiFi struggles to give the dynamic bandwidth the human hear wants. The signal needs to be compressed before recording (sort of squeezing it all onto the master tape) and expanded to its original upon playback. This tricks the ear and brain into imaging a full spectrum of sound. Equipment to do this has been around since the '40s. Sun Studios, King Records, Abbey Road studios, Beatles recordings, all were subject to this process. Recording engineers used pieces of gear such as a Fairchild 670 variable Mu compressor and expander. A big lump, 14 transformers, 20+ valves etc. It gave the sound people wanted. Getting all this onto a vinyl record takes skill, the grove width, depth, spacing is determined by how loud and quiet you want to go and, how much bass is present. Basically, the louder you go, the fatter the grove and this reduces how long the record can play for.

As computers developed, the signal processing could be managed and synthesised using digital techniques. Easier to use, cheaper and more flexible. However, the same problem exists, where errors that are introduced during the analogue to digital conversion. Many recording artists are now demanding that recordings are made using the older analogue systems. Companies are building transistor and valve compressors and expanders, based on 50 year old designs. I mentioned the Fairchild box. If you can find one, you'll pay over £20,000 for it! Also, there are companies buying up mothballed vinyl record presses, producing LPs again. Not huge quantities, but on the up. There is something fascinating about placing a record on a turntable and dropping a stylus into the groove.

I see a lot of the new "retro" equipment coming in for repair. Valve equipment that has been subject to modern day manufacturing techniques. Some of the companies have no idea about the practical issues. Valves get hot for one thing. Circuit boards are inadequate along with under rated components. Modern surface mount technology is fine, where heat isn't a problem. However, in the valve environment, components need to be soldered with their wires through a circuit board rather than simply soldered to the surface. I could go on and on. Valve guitar amps are a prime example. Many will not stand the test of time due to the way they are built. Bottom end Fenders, Laneys and Marshalls all suffer with a "Chinese build". Some companies however are producing really good gear adopting principles that their Granddad's would have used! Funny how things come around.

Chapter Three

Bring back 78 rpm records???

I wrote this article originally, prompted as a result of a listeners' request during my days as a D.J.on Valle Radio, my 50s-60s show. She wanted to hear a tune only available on a 78 rpm record. Got to thinking about how recordings have changed over the last 50.

The younger generation will have no idea what a 78 rpm record is. Ask to compare it to a CD, iPod etc, and you will be told you're having a laugh.

What's a 78 rpm record, what did you get for your money? Right. It all started at the turn of the last century. Take some shellac, ground up cotton powder, slate dust, any other filler to hand and some wax. Heat this all up, stir and avoid getting too many bubbles in the mix. Press between two 10 inch aluminum master recordings discs, drill a hole in the middle, stick a label, there you have it. The final product, a nice shiny record. This method of construction lasted 'til the 1950s. Buying one of these nice records in the '50s would cost the thick end of 7 shillings. 35 pence or about 60 cents. Not much I hear you cry, but think about what you would earn a week in the '50s. A shop worker may take home £3-5 a week. So, this was a little treasure.

What did you get for your 7 bob? You got two sides of recording with a bandwidth of about 6Khz. Nothing by today's standards. Each side lasted about 3 minutes. This maximum time, popular belief has it, was governed by the width of the records' spiral grove that worked its' way toward the centre. The grove was modulated (wobbled from side to side) to contain the music. When the needle of the pick up was rested in the grove and the record spun round, hopefully, you got your sound delivered. The record, being made from this strange concoction, was not very strong, The walls of the grove had to be spaced apart by about a thousandth on an inch. Anything less and the walls of the groves would fall apart under the pressure of the pick up. So, you soon run out of space for your spiral grove on a flat 10 inch disc. The real reason for the 3 minuet limitation, I like to think is this; The "10 inch 78 rpm standard" was invented by an American company. The average attention span of an American is 3 minutes, so.... There's your answer. Why have a record lasing longer when they would fall asleep half way through and be late for another war.

Unlike CDs, you could not drop one of these records. Well, you could but the result would be a mess. 7 bob down the drain. If you did drop a record and you had a clean break, two pieces, you may have stood a chance of sellotaping the two halves together. Putting the tape of the "B" side of course. As the disc was played, you would now hear two nice pops each time the record went round. At 78 revs per minute and lasting 3 minutes, this meant you had to suffer 500 odd clicks and pops. No good really!

On what did you play these masterpieces you ask? If you had little money, you'd have a "wind up gramophone". Mechanically spring driven turn table with a governor built in to ensure the disc went round at 78 rpm ish. After a year or so operation, it sounded like a bus engine. You had a piece of bent brass tube, a flare at one end (horn) and a pickup at the other. You screwed a needle into the pickup and let the whole thing sit in the grove on your new record. Taking the motor's brake off, after a minute of winding the handle, you were rewarded with a sort of noise, purporting to be a rendition of the original recording. Watch out for the needle. If you were lucky, you'd get two of plays of a record from one needle. The needle had a hard life, stuck in a disc flying round at 78 times a second took its' toll. The needle became unevenly worn. Continuing, the needle would fail to sit in your records' grove, eventually sliding all over the place, making a mess of the disc. Clever chappies thought taking the needle out and rotating 180 degrees, you could get another play. Wrong. Doing this was like running a chisel across the record with a nice line of swarf from the surface. Game over.

A wind up machine would have cost you £6. The springs would often break. Both caged in 6" diameter drums, a repair not to be undertaken by the faint hearted. Springs were substantial devices. Take one of these out of its' cage – suicide! The spring would fly out, covering everything in a 20 mile radius with nasty black grease. The spring had enough energy to cut your head off, take down walls of Victorian houses and slice through London busses, complete with occupants. I tried it once in Tooting.

If you were well off, a bank manager, bank robber etc. you could afford a piece of furniture called a radiogram. You had an electrically driven turntable and a wireless. You now had control over the volume! and if you were lucky, you could stack your records together to be played one after the other, on your automatic turntable. Balance half a dozen of your 7 bob records on the turntables' spindle and if God was smiling upon you that day, they would drop down, one after the other, in sequence when the pickup arm had finished playing one and moved away ready for the second. If you hadn't been to church that morning, the records would randomly drop down, on top of the pickup whilst playing. All hell would break loose. You would put your pipe down, jump up from your arm chair and run over to the radiogram to make rescue plans. Records would break, the pickup and or needle would become damaged, the mechanism would rip your arm out of its' socket, you would end up in hospital and be £3 the poorer having lost your records to this daemon machine. That's why you had a wireless built in you see, because you had no more records left to listen to.

Radiograms in the late 1930s could cost anything from £20 for a budget set, bought on the tick and do a runner after a year, to £140 for a nice McMurdo, RGD, Decca etc. In 1940s, Decca invented a full range pickup which was to revolutionise home reproduction. Sound that is, not making kids. The pickup was much lighter, had a sapphire or diamond tipped needle. The output of this was fed into a well designed valve amplifier. You only had about 6 watts of power, more than enough. Hi Fi had arrived in the UK.

CDs, I agree are much more convenient. Try playing a 78 rpm in the car going 70 Kilometres over the roundabouts through Albox. But, you had a feeling of achieving something when setting one of these to play. Bring back 78s. At least I would be able to sell my stock of needles!

Chapter four

Vinyl Records – The Come back!

I previously wrote about early recording techniques and 78 RPM records etc. The early '80s saw the CD being introduced and we all watched "Tomorrow's World" as they heralded in the new disc, saying it can even be played with jam spread on the surface! Ever tried it?? We were told the days of the LP and 45 rpm were numbered. Or were they?

The vinyl LP (Long Player) has been making a big come back. I have been supplying a steady stream of clients with restored record decks and pickup cartridges allowing mothballed collections of LPs and 45s to be played again. And, Long may it last.

The start of the Vinyl LP goes back to just before WW2. There was a bit of a battle between U.S. companies Columbia Records and RCA Victor. With improvements in recording techniques, there was a requirement to provide a recording media that was less "noisy", could play for longer and was more durable than the brittle 78 rpm which if sat on, well.... Various formats were thrown up. These ranged from 10" to 16" diameter discs. A speed of 33,1/3rd rpm was commonly used as this was convenient for gearing down a synchronous motor running off mains electric (60 cycles, in the USA). The larger 16" discs were never really aimed at the public as the players would have been massive, needing you to buy a bigger house. They tended to be used in radio stations and contained pre-recorded music and adverts. Bet you didn't know that.

After WW2, the record companies got their act together and settled on a common standard. This was a Long Playing 12" disc (LP) running at 33 and a bit rpm. The width of the groove was reduced to .001" as opposed to .003" used in 78s. This, coupled with a slower disc rotation, meant you could get up to 30+ mins. of material per side. Pickups had moved on from the big heavy magnet-coil types and now the needle that sat in the groove was weighing a few grams rather than that of a Chevy V8 engine. The plastic vinyl material had a much smoother composition over the shellac used in 78s. This reduced the background noise. The only thing that the different manufacturers didn't do until the '60s was to agree on the recording characteristics (how the sound was equalised to sound the best), which meant your HiFi enthusiast had to have a knob on his amplifier to "equalise" the playback sound of different LPs. RIAA, LP etc.. 'Spose it justified all the money spent!

RCA released the 45rpm 7" record in the late '40s. It was aimed at the popular music market, giving a play time of around 3 mins. As your average American only had an attention span of no more than 3 mins, it worked out just fine! This format came with a larger centre hole and was ideally suited for the automatic jukebox mechanisms. The increase in rpm (velocity) also meant that a louder recording could be made and replayed. This helped the sound reproduced on smaller budget players with only one or two valves in the amplifier to be louder.

10" 33rpm records were also made. These could contain 3-5 tunes per side and priced as a poor man's LP. The format lasted longer in Britain than in the U.S, being used by Skiffle, Rock n Roll bands and comedy. "Best of Sellers" being an example.

During the '50s, the EP 7" 45rpm was released. This was an extended play record. The groves were closer together, so more tunes could be squeezed onto the disc, getting around 15 mins per side. The downside was, that the music recordings had to be 'compressed' more than normal, in order for it to sound ok. This resulted in a lower volume when played. Those with a budget Dansette, struggled to make a big racket.

By the late '50s, 78rpm records were being phased out. Although recordings of Hillbilly music for those in the U.S, down south, continued for a while, as they couldn't afford a new fangled player!

It was normal for a record machine to have 4 speeds to cater for the different record types. Speeds were 78, 45, 33 and some 16 rpm. The latter slow speed was for records that were normally "talking books" and Language Learning. As the groove width for 78s and LPs (Microgroove) records was different, the pickup had 2 styluses. You would have to rotate a knob on the front of the pickup arm to select LP or 78. Playing a LP with a 78rpm stylus could be a bit messy, reducing the resale value of your LP!

Then, stereo came along...Next chapter. Bet you can't wait, really.

Chapter five

Vinyl Records in Stereo!

Last time we had a little focus on how vinyl records have been making a come-back. I said I would follow up with how vinyl stereo came about. But, not before telling you about a wine cooler I repaired. I tongue-in-cheek said to the customer I needed a bottle of Chablis in order to test it fully. The client came in with one! My happiness then to be dashed, as a Dyson vacuum came through the door.

Most of us have been blessed with two ears. As a result, to be able to listen to a truthful rendition of recorded sound, two channels need to be reproduced, for ears left and right. Stereo recording had been around since the '30s. But, getting this onto and off a record has not an easy task. Many different ideas were tried, including this one which I find amusing.

A U.S. company took a double-sided master disc and designed a head cutting machine that would cut the left channel on one side of the disc and on the other side, the right channel. However, getting the two cutting heads to start and track each other with 100% accuracy was a nightmare. The other problem was how to play the dam thing. The best solution was to play the disc vertically, i.e. standing on edge. A bit like a '50s Juke box. Two pick-up arms were counterbalanced, one set to play side 1 (left) the other set to play side 2 (right) at the same time. Unfortunately the whole mechanism was very messy. The pick-ups had to hit the lead-in groves of each side of the disc at exactly the same time. If not, you would hear left and right channels that could be adrift by one or more revolutions of the disc, making the reproduction nonsense. Eventually it was decided that, given the mechanical and technical headaches and Joe-public's cack-handidness, this idea would be binned.

Another idea was to modulate the grove cut into the record in two directions. Now in mono (one channel), the grove was modulated (wobbled) from left to right. When the record is played, the stylus (needle) would be wobbled from side to side, with a signal being created by it being physically connected to a magnet and coil and fed to an amplifier. With this new Stereo idea, the grove was modulated up and down as well. The design of the pick was such that two coils would detect both axis of movement of the stylus, producing the left and right signals to the amplifier. This sort of worked, but the main problem was distortion and interference between the two channels. The stylus in the grove would get "confused" with the up-down and left-right movement, resulting in a muddled poorly defined sound. This became much worse on loud music passages when the grove was modulated to its fullest limits.

A company called Westrex came up with a modified grove cutting head, similar to the above, but it was turned through 45 degrees. This meant there was an equal left and right channel modulation of the grove, a complex wiggle pattern of up and down, side to side if you want. The pick-up stylus was designed with coils and magnets arranged at 45 degrees to detect this signal. In fact, one coil detects the horizontal movement that produces a signal that is the sum of left + right, whilst the other coil picking up the vertical movement creates a left − right summed signal. A lateral coil in the pickup is used to offset and remove any the added signals, resulting in a good quality left and right signal with a high degree of separation.

The other benefit of this principle (if you couldn't afford a grown up stereo system), was that the record can be played using a mono pickup cartridge. This is because the stylus, connected to the signal generating parts, coil, magnet or crystal, recognised the overall movement as a sum of the left and right channels. As a result, you did not lose any of the recorded left and right signal component. It is amazing that this 45 degree cutting technique was first developed by a brilliant British engineer called Alan Blumlein, working for EMI, 30 years earlier. Alan was instrumental in developing stereo sound recording and playback on records and cinema. During WW2, he went to work on developing radar and was tragically and mysteriously killed in a plane crash whilst on a secret testing mission. His work on stereo was shelved, to be adopted eventually by Westrex, but as with lots of things like this, was never given full recognition.

So, there we have it. A record cutting principle was put forward 80 odd years ago and still used to this day! CDs eat your heart out.

Chapter six

The Humble Car Radio, bit of a change from records

Seems like cars have always had radios. Not so. Here's a true story. You'll like this, not a lot.

One evening, in 1929, two young men named William Lear and Elmer Wavering drove their girlfriends to a lookout point high above the Mississippi River town of Quincy, Illinois, to watch the sunset. A romantic night to be sure, but one of the women observed that it would be even nicer if they could listen to music in the car. Typical! Lear and Wavering liked the idea. Both men had tinkered with radios (Lear had served as a radio operator in the U.S. Navy during World War I) and it wasn't long before they were taking apart a home radio and trying to get it to work in a car. Not as easy as it sounds. Cars have ignition systems, generators, spark plugs, and other electrical equipment that generate noisy interference, almost impossible to listen to the radio with the engine running. One by one, Lear and Wavering identified and eliminated each source of electrical interference. When they finally got their radio to work, they took it to a radio convention in Chicago. There they met Paul Galvin, owner of Galvin Manufacturing Corporation. He made a product called a "battery eliminator". A device that allowed the old battery-powered radios to run on household AC current. As more homes were getting wired for electricity, radio manufacturers were making more AC-powered sets. Galvin needed a new product to manufacture. When he met Lear and Wavering at the radio convention, he found it. He believed that mass-produced, affordable car radios had the potential to become a huge business.

Lear and Wavering set up shop in Galvin's factory. When they *perfected* their first radio, they installed it in his Studebaker. Galvin went to a local banker to apply for a loan. Thinking it might sweeten the deal, he had his men install a radio in the banker's Packard. Good idea, oops. Half hour after the installation, the banker's Packard caught on fire (They didn't get the loan).

Galvin didn't give up. He drove his Studebaker nearly 800 miles to Atlantic City to show off the radio at the 1930 Radio Manufacturers Association convention. Too broke to afford a booth, he parked the car outside the convention hall and cranked up the radio so that passing conventioneers could hear it. That idea worked. He got enough orders to put the radio into production.

At today's cost, a radio for a new car would cost about $3,000. In 1930 it took two men several days to fit a car radio. The dashboard had to be taken apart so that the receiver and a single speaker could be installed and the roof had to be cut open to install the antenna. These early radios ran on their own batteries, so holes had to be cut into the floorboards to accommodate them.

The installation instructions run to 28 pages! Selling complicated car radios that cost 20% of the price of a brand-new car wasn't easy in the best of times, let alone during the Great Depression. Galvin lost money in 1930 and struggled for a couple of years.
But things picked up in 1933 when Ford began offering their radios pre-installed at the factory. In 1934 they got another boost when Galvin struck a deal with Goodrich tyre company to sell and install them in its chain of tyre stores. By then the price of the radio, installation including, had dropped to $55. The Motorola (a play on the name of the record company Victrola) car radio was off and running.

In the meantime, Galvin continued to develop new uses for car radios. In 1936, he introduced push-button tuning! Then introduced the Motorola *Police Cruiser*, a standard car radio that was factory preset to a single frequency to pick up police broadcasts.
In 1940 he developed the first hand-held two-way radio. The Handie-Talkie for the US. Army.

A lot of the communications technologies that we take for granted today were born in Motorola labs in the years that followed WW II. In 1947 they came out with the first television to sell under $200. In 1956 the company introduced the world's first pager. In 1969 they supplied the radio and television equipment that was used to televise Neil Armstrong's first steps on the Moon. 1973 they made the world's first hand-held cellular phone, one of the largest cell phone manufacturers in the world. All started with the car radio.

What ever happened to the two men who installed the first radio in Paul Galvin's car, Elmer Wavering and William Lear? They took very different paths in life. Wavering stayed with Motorola. In the 1950's he helped change the automobile experience again when he developed the first automotive alternator, replacing inefficient generators. The invention lead to such luxuries as power windows, power seats, and, eventually, air-conditioning.

Lear also continued inventing. He held more than 150 patents. Remember eight-track tape players! But what he's really famous for are his contributions to the field of aviation. He invented radio direction finders for planes, aided the invention of the autopilot, designed the first fully automatic aircraft landing system and in 1963 introduced his most famous invention of all, the Lear Jet, the world's first mass-produced, affordable business jet. Not bad for a guy who dropped out of school after the eighth grade. What ever that is.

Chapter Seven

A brief history of time, Car radios!

This chapter is a bit of a follow up on what I wrote before. Now, the radio and car were invented around about the same time, Marconi did the radio and another chap did the car. One day, in the 1930s, someone had the bright idea of putting the two together. The problem was, radios in those days were rather large. Nice big valves and requiring a 20 foot long wire aerial. Having a radio the size of a bookcase on the passenger seat of your Austin 16 was a little impractical. You couldn't drive very far either as you had to plug it into the mains.

Radio manufacturers had the task of cramming this all into the size of a shoebox to be bolted under the dashboard. First thing, valves need about 250 volts to work. To get this from a 6 or 12 volt car battery, a unit called a vibrator was used. Now now, enough of that. This device was connected to a transformer and converted the car battery's voltage to 250v. All good, but these devices buzzed like mad and made an annoying whine. Second thing was to make the set sensitive enough when using a short aerial mounted on the car. This meant you had to have one or two extra valves to boost the signal. Thirdly the set had to be loud enough to overcome road noise. Again, an extra valve or two was needed.

We now had quite a big radio to fit in the car! It was quite common for this to be in two parts. The biggest unit housing the vibrator power supply and amplifier and the smaller part housing the tuner, which would be mounted in the dash. The two connected together by a big umbilical cord. All worked well, but when switched on, the current demanded from the battery, due to 8 valves and the power supply, was something like 8 odd amps. Bit like having an extra pair of headlights on. Ok if you were driving along, but in those days a car had a dynamo to top up the battery. Unlike modern day alternators, these were useless at tick-over. You soon had a flat battery if you were stationary! Don't worry, the good lord had invented starting handles a couple of years previously.

By the `50s, Mullard had designed a series of valves that worked at car battery voltage, so we no longer required the vibrator. Transistors were used for the sound, so the sets became smaller, single units. Who remembers those lovely Radiomobile and Motorola radios with push button pre-set stations! Blaupunkt brought out the first car radio with VHF/FM. This meant you didn't loose the station when going under the bridge, but it did cost half as much as the car.

Philips came out with a great little unit that played 45 rpm records. Clever system with a spring loaded pickup. Terrific, but don't drive over a bump. The Americans came up with the 8 track player. A cassette cartridge with an endless loop of tape. It was stereo, and when the tape loop got to its' end, a piece of foil caused the tape heads to jump to the next pair of tracks and so it continued ad infinitum. This meant that the Americans didn't have to turn the tape over whilst driving, not being able to multi task.

We all know that by the '70s cassettes were built in to car radios. Then 80's saw the more expensive sets having CDs. Add to that, Dolby, MP3, RDS, SatNav and so on. Come a long way since the '30s. Fundamentally all the same though. Listen to the radio whilst stuck in traffic.

Chapter Eight

Bletchley Park and secret stuff. Three of the many people I admire from the wartime effort at Bletchley.

Gordon Welchman – Bletchley Park Genius

Anyone who has read my previous ramblings knows that I admire engineers who made a big difference to our lives, yet seem to have never made the big press. Gordon Welchman is one such man. You probably have heard about Bletchley Park, the code breaking centre for the war effort and the likes of Alan Turing. The combined efforts from this establishment, in breaking enciphered Morse codes used by the Germans in WW2, is said to have shortened the war by at least 2 years. Little is spoken about Gordon.

In the late '30s, Gordon was a Fellow of mathematics at Cambridge College. By then, it was guessed that WW2 was likely. The boss at the Government Code and Cipher School (Bletchley Park), Alistair Denniston requested that Welchman join the team which included Alan Turing, Stuart Milner-Barry and Hugh Alexander. These four chaps realised that in order to have the edge over the enemy, signal intelligence, i.e. heads-up knowledge as to what was going on by the Germans was of massive value. Collectively, they wrote to Churchill, setting out the requirements for a huge increase in resources at Bletchley. Churchill reading this, stamped the letter with "Action this day", in other words, do it now or else..

Germany had the Enigma cipher machine. This was a clever electro-mechanical system which encrypted the alphabet. It looked like a typewriter keyboard but had sets of wheels and lamps. To go into detail would fill the book up. To send a secure Morse message, the operator would set the encryption code of the day and type. Letter A would get converted to V for example and so on. The receiving machine, set to the same code would convert the letter V to A. This was used to send radio Morse to the German Air Force and land army units. As the Morse was sent via radio, anyone could listen, but without knowing the cipher code, it was rubbish.

Welchman looked at the task logically. He knew that in order to listen to what the Germans were saying, the enciphered codes had to be broken very quickly. He organised radio operators to listen to the radio Morse and try to see patterns that were repeated. He saw that there were formal communications between German radio operators, in Morse, which were in "plain" un-enciphered text. These conversations included hints as to their names, locations, military status and settings of the Enigma encryption machines. This sort of detail was incredibly useful. He saw that the cipher codes were changed at midnight every 24 hrs. This information was known as "Cribs". This meant, that if the cipher code could be broken within a few hours, the whole day's traffic could be decrypted...result, happiness. He also understood the importance of analysing the detail of the way the German radio operators worked. Their nicknames, style of Morse, their common phrases etc. This all added up to help understand what was going on. This was termed "Data Analysis", the fundamentals of which are absolutely used today.

Welchman worked with Turing who had designed the "Bomb". This was an electro-mechanical machine that could simulate the way in which the enigma machine worked. This machine would be configured using information gained by the Bletchley Park radio listening teams, Morse communications etc. The machine would run a few sentences through hundreds of thousands of possible encrypted codes. If an operator saw some plain 'German speak' coming out, the code for that day was broken and in theory, all German traffic was readable.

Code breaking machines at Bletchley Park continued to become more sophisticated and more efficient. Alan Turing's machine was very much electro-mechanical. A brilliant ex GPO engineer, Tommy Flowers, had been instrumental in designing the first programmable computer using valves. This made processing the cipher codes much quicker. The machine was called Colossus. Again, there we go, Tommy Flowers. He needs to take a bow.

After the war, Gordon went to work at the Massachusetts Institute of Technology (MIT) in the 'States, teaching ground breaking computing technology. He then joined MITRE, where his work in cipher analysis put him in the forefront of secure communications.

After retiring in the early '70s, he was retained as a consultant. He wrote a book "Hut six" which was about his time at Bletchley Park. He thought it was correct that people knew about what had gone on during the war years and how much the work there brought it to a sooner end. He was in ill health and the authorities hounded him, taking away his security status, funding and making his last few years a misery. He passed away in '85.

An unsung hero, brilliant man. And, possibly why I am writing in English, not German.

TOMMY FLOWERS... More

Code Breaking at Bletchley Park

Last section, I wrote about Gordon Welchman. Unsung, and who was instrumental for some of the most incredible code breaking work at Bletchley Park during WW2. Judging by the queues of people lining up outside the shop wanting a signed copy, I thought I would remember another brilliant chap; **Tommy Flowers**. Tommy was responsible for the creation of the first programmable computer, which was key in the decryption of German Teleprinter traffic toward the end of the war.

After his apprenticeship and obtaining a Degree in Electrical Engineering, Tommy went to work for the GPO's (General Post Office) cutting edge research centre at Dollis Hill in North London. He worked on the idea of creating an all-electronic telephone exchange that could be programmed. Up until then, the Strowger exchanges were electro-mechanical, slow, high maintenance and space hungry. By '39 Flowers had put forward plans for the modern exchange.

Alan Turing, working at Bletchley Park, had been responsible for creating the mechanical "Bombe" deciphering machine. He had heard about Flowers working with electronic switching and asked his boss at Dollis Hill if he could come up with a system to help decode some of the data produced by the Bombes. Tommy put forward his ideas. However, the project never took off. Instead, Turing, being so impressed by Tommy's work, asked him to set about making a machine to decipher the German Lorenz code. This cryptographic system was much more complex than Enigma and was used in conjunction with the teletypewriter system. The magical solution put forward was a piece of electronics containing something in the order of 1500 valves! This actually frightened staff at Bletchley, including Gordon Welchman, who said it would be unreliable due to the large number of valves. Gordon favoured a mechanical version. This however, would only work at a tenth of the speed. Flowers' view was that valves were reliable if they were never switched off, which of course is spot on.

With the blessing of his boss and team at Dollis Hill, Flowers (funding a significant part of the work from his own pocket to prove a point) in 11 months, had created this monster valve machine. Put to work with cipher data from Bletchley, the system produced vital data from day one. The machine was called Colossus, for obvious reasons! Seeing the requirement for more deciphering and faster speeds, a Mk2 machine, (with upwards of 2,400 valves) was designed, installed and up and running at Bletchley by 1st June '44. This immediately started to produce crucial information relating to the pending D-Day landings. Flowers later recalled the atmosphere when a motorcycle courier presented Dwight D. Eisenhower with a pile of Colossus decrypts just before D-Day started. The information was instrumental in the planning. Without this data, things would have been very different.

By the end of the war, 10 Colossi were in service at Bletchley. As a lover of all things valves, this must have been a great sight! Just working out the power required to feed this hungry lot, would give rise to losing huge amounts of heat alone, on the basis that each valve would dissipate a couple of Watts! By '45, the 11th machine was moved to GCHQ. It was put to use in cold War activities, monitoring electronic communication traffic from those naughty Ruskies. It was switched off in '60.

Post war, the government recognised his personal investment in creating Colossus. They gave him a brown envelope with a Grand in it. Great ah? In fact, it was nowhere near what had been spent and he shared it with staff that had stuck by him. He wanted to create a computing system for commercial use, but could not get funding by the bank as they couldn't see how such a machine would work. Ironically, he couldn't say "look mate, I have made loads of these" as he was governed by the Official Secrets Act. How ironic? He continued his work at the GPO research establishment and put in place the first electronic telephone exchange installed at Highgate Wood in the early '50s. Latterly, he worked for Standard Telephones and Cables where he headed a team working on modulation techniques for sending voice calls around the world.

His work on Colossus and related equipment, set the rules for modern day computing. It was all driven with a common clock signal for accurate synchronization of all different parts, the backbone of any computer today. He was never fully recognised for this, again as a result of the Secrets Act. So, if you wouldn't mind, sit up, a quick salute to Tommy please.

Bletchley ParkCodebreaker
William "Bill" Tutte

If you have read the previous passages, you will probably appreciate that I have a passion for vintage electronics, engineers who have stamped their mark on our lives and who have gone somewhat unsung. I have written about some of the heroes of the WW2 code breaking establishment Bletchley Park, Tommy Flowers and Gordon Welchman. Bill Tutte is another good example.

During the first few years of WW2, B.P. concentrated on breaking the encryption codes generated by the Enigma machine. This was used to encode the German Morse Code sent via H.F. radio. The

Germans realised that using this system was a slow and a man-intensive way of communicating. At the sending end, you had the machine operator, someone to read out the message to be sent and a radio guy. A similar set up at the receiving end. 6 men all-in-all to send a message.

The German High Command, around '41, started to use the recently invented Teleprinter machine to communicate with strategic outposts around Europe. The Teleprinter was similar to an electric typewriter. Each character, key, pressed was sent and received as a unique "set" of closely timed 5 pulses. This was known as a Baudot code. These machines could send and receive 100s of characters a minute, far quicker than Morse and using one operator. This signal was sent via a radio network to the various German Command centers around occupied Europe. Unlike Morse, each set of characters were used to frequency modulate the radio signal. Bit like our early fax machines. To scramble this signal, a machine called a Lorenz SZ40 was used. This was like an Enigma machine but on Speed and fed Cocaine. B.P. nicknamed this Tunny. The encoded message was generated in the form of a paper strip where the holes representing the characters were punched. A fast and very secure way of encrypting this signal. Or was it?

Bletchley Park started hearing this new radio signal and soon realized it was a sophisticated automatic system. A senior codebreaker at B.P. John Tiltman, a cipher genius, soon worked out that the transmission was encoded Teleprinter characters and was based on the Vernman cipher system. B.P had acquired a Lorenz machine, but without knowing the initial settings of the machine prior to sending a message, cracking the code was impossible. Tiltman did work out that each new message was proceeded by a 12 character set of information, details as to how to set the code of the machine for that day's work. This tied up with the 12 encoding wheels within the Lorenz SZ40. Luck struck B.P. one day when two transmissions were intercepted, one after the other. The receiving station sent a reply message to the sender saying "Nicht bloody gut", German for no good. The sender committed a sin which is said to have shortened the war. He resent the message, without changing the wheel settings (encryption code) on the Lorenz machine. From a mathematical point of view, this gave B.P. a huge insight as to how the Lorenz worked. This was known as a "Fish".

The job of working out the encryption platform was given to a brilliant mathematician called Bill Tutte. Cutting a long story short, saving this magazines' paper, Tutte visualized a reoccurring pattern in the way the scrambled message appeared. This gave him an idea that there was a flaw in the way the Lorenz worked. All he had to do now was to number crunch, broadly speaking. But this needed computer power. B.P had developed a machine called Robinson (named after the mad inventor Heath Robinson), which could be programmed to sort through and apply encryption codes to received messages. It worked by reading the paper tapes generated by the Teleprinters. However, it was unreliable, used electromechanical relays and slow for what was now required of it. Tutte started to work closely with a G.P.O engineer, Tommy Flowers (I talk about hin elsewhere in this book). Flowers knew the only way to create a machine fast enough and reliable enough was to use valves in place of slow relays. Tutte and Flowers had a real battle with their superiors at B.P. Flowers was told that valves were not reliable and as there was a war on, they were in short supply. This didn't stop Flowers and Tutte. In fact Flowers spent his own money on second hand valves to create the first programmable computer (over 1500 valves in all!). Once set up with Tutte's encryption formulas let's say and after many hours of tweaking, *Colossus* as it became known, started to reliably decipher coded transmissions. By today's standard, the machine was fast, reading paper tapes and 1000 characters a second.

The decrypted information was used carefully so as to not alert the Germans as to their High Command Teleprinter traffic being read. I guess there was a fundamental flaw the German's thinking, in that they believed that the Lorenz machine could not be cracked. Hitler's world did not allow free thinkers, people whose minds operated in ways that didn't comply and so on. The sort of people that made up B.P. Hitler thought the Lorenz machine was *it* and that was that. I have very much over simplified this snapshot of technology!

It is said that Tutte and his team at B.P. probably shortened WW2 by 2 years, saving over 20 million souls. Were these guys ever recognized by the Great Britain? No, not really. Tutte went to work at a couple of Universities in Canada where he *was* recognized and received awards for advances in statistical maths. Flowers was given £1,000 to cover his cost in buying valves because funding was not available at the time. Didn't cover the costs at all. He did get an O.B.E. and a road, in his London East End, named after him, but that was in 1990. 45 years after the war! It is awful that many of the B.P. workers struggled in later life. You were shackled by the Official Secrets Act. If you went for a job interview you couldn't even mention the clever stuff you had done.

So I say thanks to all those at B.P. who worked behind closed doors, giving us the freedom we have today and saving countless lives. I will also thank my Granddad, George Parker. He was fireman in the war, he saved lives too! And, he enthused me in electronics.

Chapter Nine Coming up to date for a while.

Who is listening in?

I was talking to a customer one day about electronic eavesdropping. Strange topic I know.

Thinking about how things have changed. After WW2, throughout the Cold and War and later, electronic "overhearing" was a massive business, on all sides. All the Embassies in most countries communicated by telephone, HF radio (which covered long distances), Telex and Facsimile. All these technologies were easy to eavesdrop on. Certain departments had teams of G.P.O engineers tapping into known important phone lines, recording 1000s of hours of calls onto reels of tape for close scrutiny. Radio could be listened into anywhere. The frequencies used by Embassy officials were well known on all sides and were monitored 24hrs a day. Telex was a system of sending strings of simple text data over modified phone lines and so could be intercepted at the Telephone Exchange. Early Facsimile used a similar system.

So, how did we protect ourselves? Well all the signals used in the above could be encrypted. In its' simplest form, encryption or cipher, is a mathematical sequence of changes (algorithm), whose pattern is only known by the two communicating parties. This sequence could be changed by applying a different "key" or set of rules. Let's take the telephone. We have all heard the term "switch in the scrambler". This was a system that chopped up the audio of the phone call into its various frequency blocks. These were all messed up in accordance with the rules of the cipher code algorithm. The resulting audio was then sent over the phone line to the receiving end. Anyone listening into the call would hear a load of garble. However, at the other end, the telephone had a descrambler driven by the same cipher rules as the sending end. The blocks of audio would be reassembled in the correct order and presented as intelligible speech. A similar process was adapted to early facsimile, but by nature of the way it worked, the system was more complex.

Encrypting Telex was easier. These machines talked by using short bursts of "characters", made up from strings of 5 or 8 bits of data, electrical on-offs. Each letter had its own unique data pattern. So, all you had to do was to apply a set of rules that said, "convert letter A to D, S to U" and so on. At the other end, apply the same rules and out pops the original text.

Now comes the interesting bit. Let's say you wanted to monitor a company's Telex machine, but you couldn't listen in at the Exchange and you didn't have the Encryption Cipher. These machines printed the text by using small solenoids to punch ink onto the paper. Some had monitor screens as well. With not too much in the way of techie gear, sit yourself in a van or building near by with a sensitive directional aerial aimed at the building, coupled to a modified radio receiver. This would detect the spikes of energy given off by the Telex machines mechanism. You could also receive the signals being generated by the monitor screens. With all this recorded, back at the "lab", with some jiggery pokerey, one could easily replicate the original data sent.

The US and British government introduced a standard that machines used for secure data had to meet to prevent the possibility of "electronic listening in". This dictated that the electrical emissions of the machine had to fall below certain levels before they would be certified. The standard was known as *Tempest*.

Listening into conversations in rooms was easy. Small radio transmitters fitted into plug sockets that transmitted low range to the listening party. Or, how about using an infra red beam, aimed at a window of an office from afar? One could detect the small changes in the reflected beam caused by vibrations in the glass as a result of the voices inside the office. A bit of filtering and amplification, there you have it. Crystal clear!

Modern telephony, mobiles, internet etc... makes a very easy way for the authorities to monitor us. And, they do. More on that in the next publication. Look over your shoulder. Be careful.

Who is listening in part 2

In the beginning of this chapter, I touched on the topic of electronic eavesdropping and so on, post WW2 and into the '70s.

In those years, the listening-in was mainly related to 'phone lines, telex and latterly fax, along with radio transmissions to and from Foreign Embassies. This was well established and was understood on a gentlemanly basis.

As the Internet developed in the '80s and beyond, with emails and on line shopping, a new breed of eavesdropping developed. Most of us at some point have been on the receiving end of email hacking and having various on line accounts broken into with all the inconvenience and financial implications that brings. That's a topic that could be the subject of many a rant. What is a little more worrying for me, is the how the "authorities" have full knowledge of what we are up to.

Take a simple thing like browsing the old National Interweb. You may use convenient search engines provided by, let us say, Google and Yahoo. They are free. But why? Every time you search a topic, all the information you enter is recorded and stored in their huge databases. This can include information relating to your computer, where you are communicating from and lots of other data that can be "sniffed" from your

PC. All this data can be sold on to marketing companies to enable them to target you with sales dross, hopefully which may be of relevance to you, given your past history of web searches.

So, you get a nice email advert, tempting you to "click" on here and there to get a fancy Search Tool bar for example, to make your browsing life so much easier. These can be quite subtle teases. You click and say yes. Then, you find that your web browser no longer looks like it did. All kinds of "Pop-Ups" appear when you don't want them, offering all kinds…. A real annoying one of many, is "MySearchDial". This can modify your Internet Explorer, Chrome or Firefox, so that when you open them, you are taken to a marketing site which records an awful lot more than you think. If you see this on your machine, or you find internet browsing is dog slow, call me. I will tell you how to remove it. The old adage, be careful of things that look too good to be true! Also be aware of "Make your PC Faster, Tune UP, Free PC Checkup" etc… These things are free and you wonder why. Most of these Trojans place programs in your PC to watch and report what you do.

If that has worried you, what about sending an ordinary email? Well most Internet providers are obliged to make data you send available to big bro. Your emails will all be parsed / filtered through some clever software, looking for certain keywords, phrases, names etc. e.g. that could be terrorist related. Along with your address, account and recipient and senders details, all gets stored. If certain criteria are met, it will be subject to a more in-depth manual search. You may have seen some companies whose email systems reject automatically rude words.

Scunthorpe being an example! A similar process. Ah, you may say, I will encrypt all my emails. And you can. But watch out. You are not allowed to use an encryption system that is not government approved. That's because they need to be able to crack it! Doing this may lead to an unwelcome knock on the door, an injection of Sodium Pentothal and your toenails being removed.

Telephone calls, both mobile and landline come under the same scrutiny. Most mobile calls are recorded and again the speech is cleverly analyzed to detect phrases, names and whatever happens to be "hot" at the time. Same for SMS. With your mobile in the pocket, your approximate location can be worked out should it be needed, as the 'phone company if asked, must provide your cell site and log on details. Slightly more concerning is that with most mobile devices, iphones, ipads and so on with GPS enabled, the operator, in theory, could read your location, down to a couple of metres.

There we have it. Just a little more on the subject of big brother watching you. If you want to communicate safely, I suggest rice paper, invisible ink and an exchange of brown envelopes on platform 3.

Chapter 10

All our eggs in a Chinese basket!

We all love our electronic techie toys, phones, PCs, labour saving gadgets, TVs and so on. All now relatively cheap compared with 50 years ago. Remember in '67 a 22" colour TV was a similar price to that of the Ford Anglia van that delivered it to your door. As a kid, I had a transistor radio which was stamped "Empire made" i.e. made in Hong Kong.
Nowadays, a very significant percentage of our electronic consumables are made in China. Mass production and very low labour costs combined with high work ethics have driven this on.

All good, but there is a big worry. We have become very beholden to the Chinese manufacturers. I say we, I mean the Western World. Let's have a look at one of the biggest global manufactures of communications equipment, Huawei. You may have seen the name on your clone of an iPhone for example. I am not picking on this organisation, just making an example. Huawei started life as a reverse engineering company. It took existing *Western* products, pulled them to bits from physical and software aspects, with a view to reproducing similar and cheaper clones. Now, the company is one of the biggest telecoms and IT companies on Gods' planet. They make, routers, fiber transmission systems for the telecoms networks, you name it. In the early 2000s, there was a directive that BT (British Telecom) should convert all the old analogue-digital Time Division multiplex voice networks to the latest I.P. (internet protocol) systems, in readiness for the data explosion. In order for this to be achieved, millions of pounds had to be invested in new gear. After much debate, the best value for money came from Huawei, sending it's competitors such as Marconi to the wall.

Is it a wise thing to entrust your entire communications backbone, with all it's implications such as national security etc. to one company? **No** was the blunt answer. An organisation was set up to closely monitor and report back to the government on foreign made data equipment. A highly secure establishment was set up known as the "Banbury Cell", real name Huawei Cyber Security Evaluation Centre (HCSEC). Yep, heavily funded by Huawei! Top Bletchley Park type IT geeks and scientists pull apart hardware and software coding that makes up all the 'comms equipment and reports on and advises remedies on vulnerabilities. This to placate UK stakeholders and politicians. It seems a bit silly to me that the manufacturer was allowed to monitor itself to a large degree. In fact things have now changed in that GCHQ and the older CESG (Communications Electronic Security Group), now combined into

GCHQ, have taken a significant control over the *Cell*. Vetting of staff and procedures are monitored closely. Why is all of this such a worry you ask? Well, consider a time of crisis, not even a pending war, some sort of financial crash or global situation. If an interested party had the ability to monitor a Nation's data and voice traffic and manipulate it at will, by secretly listening and filtering off the key good bits, to whatever advantage, that could be very attractive. With the lion's share of our communications in the hands of a potential enemy, the results would be catastrophic. In fact in 2011 a "backdoor" opening was found in a piece of Huawei equipment. This would have allowed those technically minded, to access the inner working of a router and sniff around any data being carried. An engineer had left some access port used for remote programming, accidently open. My point exactly. That engineer is now probably planting rice in a paddy field. Strangely shortly after, Huawei invested 650 million quid into their UK operation. This was exactly the same sum that Cameron pumped into the Cyber Security establishments. Call me a cynical old git, but was there a link?

See, there is a club called the 5 Eyes. This is made up of experts from Australia, US, UK, Canada, New Zealand. They pool all their security and technical resources, share it and watch out for each other's back over a beer and a BBQ. I like the Aussies. They looked at buying kit from Huawei and said "Fxxx that cobber". They were too worried about exposing themselves. So to speak.

I must add that it is not just Chinese made equipment, although this issue is now making big news as governments are looking at rolling out big 5G mobile data and voice networks. Huawei is a main supply contender. Similar systems are made in Israel, UK and US, all of which should come under the same scrutiny in my books. Be warned, if you Google any of the above, watch out for a black van with a spinning roof rack parked outside your house.

Chapter 11

The NHS RansomWware virus

Well, what the hell was that all about? We recently saw how the NHS and many other companies and computer users were brought to their feet by a nasty PC virus. One small piece of computer code caused so much damage. And it doesn't stop there.

What was behind the attack? It goes back a few months before it happened. In previous articles I wrote about Internet Hacking. The U.S. had commissioned a series of viruses (worms) back in around 2010, which could be used to wreak havoc with Iran's nuclear production process. This bit of software, when loaded into industrial control systems would attack standard off-the-shelf Web-Enabled motor controller units. In the Iranian situation, this upset motor control systems that were used in centrifuges refining nuclear material. They went out of control and damaged themselves. So, what does this have in common with the recent disaster?

It 's well accepted that the NSA (National Security Agency, U.S.) had designed a virus, EternalBlue, which took advantage of a weak spot in Microsoft operating system. It was a problem with how M.S. dealt with remote server access. Basically, if you had the where-with-all, you could access and walk around any PC on a network. A disgruntled ex-employee had collected these Trojans and viruses, including EternalBlue and had them on the Internet, up for the highest bidder. For whatever reason (he probably had the hump, his XBox broke) one day, he just let access to be had by all. He was arrested. In March, Microsoft had issued updates for systems like Vista through to Win10, to shut off this loop hole. The damage was done though.

A group know as WannaCry took a variation of EternalBlue and added some nasty additional payloads. The virus, hiding as an update attachment such as Adobe etc. Once on your machine, it did several things. The first was to "open" your PC in readiness for remote access via a third party, should it be needed. The second, was to install an encryption package to make your files unreadable, unless you purchased a code. Thirdly, the virus locked your machine so you could not carry out any repair applications. Lastly, it looked at any PC connected on your network and tried to spread itself to carry out more of the same crimes. Once infected, your screen showed a page where you could pay to have your machine unlocked. The payment was in the electronic currency, Bitcoin. 12th of May, the virus was released.

Within a few hours, AntiVirus companies were seeing reports of networks globally being damaged. The High profile NHS hit the news as 60% of their network was compromised. Many trusts had not updated their systems. Planned operations, to local Drs. Surgeries were shut down, with an inability to see patient notes. AntiVirus companies, GCHQ etc., were looking at ways of stopping the spread. Strangely enough, it was a chap who worked at an AntiVirus company, monitoring what was happening, he was on holiday. He managed to load the virus onto a sacrificial machine, so he could see what it was doing. Inadvertently, he noticed that the virus was trying to make contact with a website that didn't exist. This may have been an oversight on the Hacker's behalf. He bought the domain name for a few quid. Long and short, this stalled the virus spread and allowed remedial work to be done. One chap, on holiday! The government recognised his efforts and gave him 10,000 quid and some extra cheese on his pizza. He gave it to charity. The money that is, he ate the cheese.

So what does it all mean? We are all vulnerable. We live and die by the Internet. Just imagine if you took that away? No Facebook, emails, TV and telephone calls. Very frightening?

Think of this I ask you. WiFi connected equipment is all common place. You can control your house heating, lighting systems from your tablet, evens children's toys can be accessed from a laptop. There are many easily obtained programs that you can download which will enable you to "sniff" the WiFi air and hack away. Knowing the right things to do and assuming default passwords such as 1234, you could have access to your heating system. OK, no big deal you might say. But, if Dodgy Bob hacked into every one of these heating controllers across the UK, bought at B&Q, and decided to whack them up full 12.00 hours on Christmas day, the Energy Companies would not cope with the overload. **SHUTDOWN**. No sprouts, no turkey.

Chapter 12 Back in time again

T.I.M. Dial 123 for the speaking clock!

In the next chapter, I write about a chap caller Mr. Strowger, inventor of the automatic telephone exchange. He set the way to enable us to call other phone users, without having to go through the operator, asking to be "connected" please. But you'll have to read this first.

Up until the mid '30s, different areas of the U.K. had their own independently run telephone network. This mismatch lead the government to bring everything together under the control of the G.P.O. (general post office).

The cost of rolling out the telephone network was costly and the G.P.O entered into a massive advertising campaign to sell the service. One idea to show off this new technology was to provide an accurate speaking clock time service. The G.P.O research laboratory at Dollis Hill, North London, set about creating the speaking clock machine. The result was a masterpiece of electro-mechanical, opto-electronics. Basically, the machine had 4 glass discs, about the size of a 7" single. These discs contained recordings of words, in a similar way as audio was recorded into the edge of cinema film, sort of black and clear troughs of transparency. As the information was sealed in glass, it would never deteriorate. The discs were linked on a shaft, rotated by a highly accurate speed motor. 2 discs had spoken minutes recorded, odds and evens. The other 2 discs had hours and other words. A set of special lamps, lenses and photo cells read the audio from the discs. The position of the photo cells determined which words were read. Every 30 seconds, the cells, moved to different parts of each discs via an intricate cam system. So, every 30 seconds a complete sentence would be reproduced by playing parts of each disc one after the other. "At the 3rd stroke, it will be 10, 23 precisely". 80 odd different words had to be stored. The speed of the main motor controlling all the actions was locked to a central signal, governed by the Greenwich Observatory. The accuracy was about 10th of a second! The first recorded voice used was that of a Miss Caine. She had won the competition (15,000 entries) set by the G.P.O. She became known as the girl with the golden voice. The recording of the phrases was done in one take and took a couple of hours. The machine went into service in High Holborn Exchange in 1936. In fact 2 machines were fitted. Should one fail, the other took over automatically. They were both synchronised together. With the associated electronics, it fitted into a large front room. Dialing 8,4,6, (TIM) connected you to a distribution system, feeding you the speaking clock. Initially the service could only be heard around the London area. A few years later, the same system was installed in a Liverpool exchange. So you could now accurately tell when your car wheels were nicked. With the creation of Subscriber Trunk Dialing (STD, no, not a nasty transmitted disease), whereby you could call any phone user without operator intervention, the number for the speaking clock became 1,2,3.

The public's need for accurate time telling took the G.P.O a bit by surprise. In the first year of service, over 20 million people called the speaking clock. The revenue as a result was about £85,000. Between 1936 to '49, Nationwide, 400 million calls were made and 40,000 called between 8 am to zero hour on Armistice day. During the war years, the speaking clock system was made ready to broadcast recorded messages and instruction to the nation should it have been needed.

Throughout the following decades, different voices were used, sponsors such as Accurist came on board. Machines used magnetically stored speech and ultimately became a solid state computer controlled system the size of a desktop PC.

Writing about this made me remember how your telephone bill was calculated. In the old electro-mechanical Strowger exchanges, every subscriber (landline user) had their own Fee Meter. This was a device that read fee meter pulses that were superimposed on your line, when a call was connected, in such a way that you could not hear them. The meter did however. The number of pulses per minute depended on time of day. At the end of each month, a photo was taken of your meter and the units compared with that of the last photo. A lady would punch the numbers into a comptometer and a bill to pay would follow. How quaint!

Chapter Thirteen Still back in time

Mr. Almon Strowger. His automatic telephone exchange.

And here is another chap that gave us something interesting to play with.

We all take ringing someone up for granted. It wasn't long ago, 1920s, that you would pick up the phone, to be answered in person by a polite lady, asking with whom you would like to be connected. Picking up your receiver would operate an electro mechanical flap, "Dolls Eye" at the local telephone exchange to attract the attention of the "operator". She would ask you who you wanted to speak with. Using a jack lead, she would connect your line socket to the other person's socket, wind a generator handle to make the distant end telephone ring. She would observe the

"Dolls Eye" and when it flipped back, as you put your phone down, she unplugged you.

Now, history has it, that Mr. Strowger, an Undertaker by trade, had the right hump with the girl at his local telephone exchange. He knew she listened in on his calls and passed pending business to her husband, a competing Undertaker. Awful. He set about designing the principles of an automatic system which could connect telephone users together, without the human element. He enlisted the help of his brother-in-law who was an electrical engineer. A system of pegs set in a circle were connected to a central relay system. A specific sequence of peg operation connected you to the required other user. This system, that now had a patent, was installed in Strowger's then home town in India, in 1891. "The Automatic Electric Exchange Company" was established. Strowger sold his shares and patents in 1898 for about $2000. When it was realised how clever the invention was, they sold for $2.4 million in 1916! Bugger!

The system developed quickly and in the UK, the British Post Office trialled several systems around the country from different manufacturers, from Sweden, Germany and the US, Mr. Strowger's, won.

The system developed into what's known as the "two motion" switch. You will be familiar with the dial telephone. Lifting the handset enables a relay system in the exchange to get ready for your call. As you wound up the spring-loaded dial and released it, it would wind back at a regulated speed, sending a chain of pulses (10 per second) to the exchange. 1 pulse for digit 1, through to 10 pulses for digit 0. This would step a "selector relay" (bit like a 10 way light switch) setting up the first path to your required called party. With each wind of the dial and each digit dialled, a different "selector" would take you to the next part of the route. Eventually, when all digits were dialled, a "ringing current" would be sent to the distant phone, to ring its bell. The subscriber would pick up the handset, which would finalise the electrical connection between caller and receiver at the exchange. The answering party would politely announce themselves as the name of the local exchange and your number. E.g. "Temple Bar 3658". Or, if in Spain, an abrupt "digame". The exchange would detect if the distant handset was ready to receive a call and not busy. If it was in use, a busy tone would be sent to the caller.

All the tones you would hear, i.e. Dial tone, Busy tone and the individual words making up the talking clock were all recorded on a huge magnetic rotating drum, driven by a large motor at constant speed.

The electro-mechanical Strowger system stayed in service in the UK until well into the 1980s. As the number of simultaneous calls were limited by the number of cables between exchanges, in "Times of Crisis", telephone access would be restricted to police and doctors (known as Class Of Service). If you picked up your handset during busy periods, you may have heard "All lines are busy user".

Interesting fact... The Strowger, being an electro-mechanical system is far less prone to damage by an "EMP", Electro Magnetic Pulse, generated by a nuclear bomb. The EMP would trash any modern day electronic system. An old Strowger system, under Downing Street, is currently maintained, just for that reason. Top Secret! If I told you any more, I would have to shoot you.

It remains for me to say, "Press button B to get your money back". If you understand what that means, like me, you are too old.

Chapter Fourteen Yep, still back in time

Home Chain Radar

This is fun. Well, before World War II started, the government set out to provide an early warning system which would be able to detect nasty airborne fighters before they hit the shores of Blighty. Having a fondness of vintage electronics, the technology and equipment used initially was quite primitive, but lead the way to far greater things.

The project to sort this out fell in the lap of Sir Robert Watson-Watt, clever chap, who was the technical head to the British Air Ministry. He came up with a plan which was to be the first fully "linked" radar station network. This was to, in principle, put a chain of radar stations all around the coast line of the UK, giving a contiguous radar view out to sea. Starting along the South East, the stations looked toward France where the airborne *Luftwaffe* advance was likely to come from.

As said, by today's standards, the technology was very basic. In fact by 1930-1940s standards, the Home Chain radar was thought to be outdated. But Watson-Watt stood his ground. Churchill wanted something **now**. The technology was to build 350 foot tall masts with aerial arrays slung between them. These would be fed by powerful HF radio transmitters using the 12 mtrs, 25 megacycle frequency range. These arrays transmitted a signal out across the channel. A few hundred yards away would be another mast and aerial setup which received any small signal reflected by potential evil aircraft. Yes, the systems were basic, but the manufacture of the equipment would be easily taken on by domestic radio companies such as ECKO, HMV and EMI, as well as Cossor, whose labour force was now diverted to the war effort.

The receiving huts were manned, or "womaned", normally, by WAAFs, Women's Auxiliary Air Force. These nice ladies, dressed in khaki uniforms, stockings…sorry I must stop there, would huddle around a small TV like screen, about 6" diameter. The image comprised of a splodge at the bottom of a horizontal line, which indicated the stations land position. The rest of the screen had lines, marking distance in miles away. The reflected signal from an aircraft was seen on the screen as another ill defined mass, X number of miles away. Of course this could have been a massive flock of birds or heavy cloud. But a trained WAAF could distinguish what was what. This information was collated and telephoned through to RAF Fighter Command at Bently Priory. No email and *WhatsApp* in those days. There, the "Map Plotters" would determine the plan of attack and send commands to scramble. The information also went to the Navy and the Barrage Balloon controllers.

The system did have limitations. The radar didn't have spinning dishes like you see on boats. It was fixed and could only "see" with about a 60 degree view. That's why many stations were chained together, to give an end-to end view along the coast. It also wasn't much good at detecting low level aircraft. The system known as AMES1 (Air Ministry Experimental Station) was upgraded throughout the war. AMES2, 13... was developed using higher frequency systems which increased accuracy and could see aircraft coming in low level, hugging the sea.

By 1943-44, new generations of radar were being developed and deployed, an example being Ground Control Intercept radar. This used Microwave frequencies, 1.3 gigahertz. The power being generated by a newly designed chunky looking valve known as a Magnetron, similar to the device used in your microwave oven. The images and detail were far better than the early systems. This now looked very much like the radar systems that are used today. Airborne radar, i.e. smaller versions of the land based systems were being installed in aircraft. Much of this development work was done by EMI's team, headed by a brilliant engineer called Alan Blumleine. Sadly and with a cloud hanging over the whole matter to this day, his career was cut short when an aircraft carrying Alan and several other top engineers mysteriously crashed while making secret tests.

So there we have it. A system what was very crude and ridiculed at the time by many, but worked and did what it said on the tin. That phrase would not have been understood in 1940! So please say thanks to Sir Robert Watson-Watt when you get a minute.

Chapter Fifteen Back up to date

A brief history of the Internet! Two parts

-

We have been digitally communicating with each other for well over a hundred years. We had Semaphore Watch Towers, and Railway Telegraphy where electromagnets, energized from far down the line, that would move indicators in signal boxes.

Our modern day Internet goes back to the late '50s. As a result of those naughty Russians popping up the first satellite, Sputnik, the US became even more paranoid. The Cold War was at its height and President Eisenhower wanted a communications network that could withstand THE BOMB. The

Advanced Research Project Agency, ARPA, was set up.

In the mid '60s, a principle engineer at ARPA, Leonard Kleinrock, set up a network in which various computers, in different locations in the US, were connected together in a sort of matrix. The clever bit here was that the computers could carry on communicating amongst themselves, even if one or more of the links were cut (e.g. by a nuclear strike). This set the corner stone by which today's internet was built.

Kleinrock continued to develop mathematical models where data would be transmitted between computers in so called "Packets". A protocol called TCP/IP (Transmission Control Protocol / Internet Protocol) was born... ARPAnet. This is what we now use on a daily basis. It sort of goes like this: Everything we do on the internet, be it Skype calls, video streaming, emails, pictures etc. is all converted into data. A series of long streams of logical 1s and 0s (Bits), electrical on and offs. These streams of data are packaged up into convenient "Packets" of data for easy handling. Now each packet of data contains some additional information, such as an address, the order in which packet was made and the mathematical sum (Checksum) of all the Bits inside it added up. It's a bit like a series of envelopes containing one word of a long letter being sent to a loved one.

Let's say you want to send an email. The text, attached photos etc. are packaged by your PC and sent into this massive postage system called the internet. All the packets have addresses, so at some point will arrive at the destination PC. The Internet being this massive web of connections (making it resilient to parts being damaged), your packets could take one or more of millions of routes, depending on which path is available at that moment in time. Parts of your email could get to the destination via routes from, the US, India, Spain, UK and so on. They also will arrive at slightly different times and not in order! However, because each packet has an address, an order number, your PC will arrange everything correctly, leaving the email to make sense. If there are any errors in the packets, which the receiving machine will suss out because the checksum will be wrong, the receiving machine will politely ask the sending machine to resend that dodgy packet. As mentioned, even our speech (Skype for example) is sent as packets of data. When the speech goes all *Darlek* type, it's normally due to poor local internet capacity, where packets are lost and or take too long to reassemble in order, making speech unintelligible. Remember, all this happens in the space of a few hundredths of a second! There you have it. Easy.

Whilst on the subject of the Internet, if you are struggling with your download speeds, buffering TV, poor Skype etc. speak to me about an alternative Internet system. Telplay is now becoming a serious contender in this area. Well established elsewhere in Spain, they have an independent network and guarantee the bandwidth you subscribe to, any time of the day or night. Talking to their engineers, it's refreshing to see that they will not add users to a point where the network is overloaded and you end up contending for bandwidth.

Internet@war.....

A while ago, I wrote about how we could all be subject to electronic eavesdropping and internet monitoring. That was relating, more or less to the public end user. When it comes to Government, Military and Corporate matters, that is frighteningly different.

A huge percentage of data communications used by banks, Military and infrastructure-control to name but a few, is carried out over the internet and therefore open to being hijacked, "sniffed" copied, modified, etc... As technology moves on, the sophistication of these threats increases at a commensurate rate.

The attacks of 9/11, started a massive review of the 'States and global Internet Security. The US Director of National Intelligence, Mike McConnel, sat with George Bush for 15 minutes, post attacks, and bluntly told him this was just the beginning. He coldly explained that whilst flying planes into buildings was bad enough, what would happen as a result of the following: Mike explained that, hacking into the systems of one or two international banks, monies could be transferred, lost, redirected. Accounts could be deleted and it would be almost impossible to repair the damage. Confidence in the money market would vanish and a financial crash would occur. He went on. The National Power Transmission is controlled via the internet. Say you were to hack into that, taking power stations off the grid and plunging numerous states into darkness? This would make 9/11 pale into insignificance. To his credit Bush was outraged and set about a massive review and amplification of USA's cyber protection.

Countries and Continents hacking into and protecting each others secure networks is a multi billion dollar business. The US spends some 3-4% of GDP on it. China is at the forefront of military and commercial attacking. The US claim that the majority of attacks on its commercial and military networks come from the Far East. In 2012 the US had some 1500 personnel with a budget of 350 million $ defending itself. However, it is said China has more that 10 times that resource. Let's take a simple fact; Big companies such as Cisco as an example, who provide data routers which allow us to connect clusters of PCs and servers to the internet, have their products made in China. It is well documented that some manufacturer's routers have been found with "backdoor access" imbedded into their software. This means that a PC with a suitable Trojan downloaded onto it, could secretly open the router to sending personal, commercial data to locations undesired, without the user knowing.

The writing of viruses and Trojans is big business. The government agencies such as GCHQ and the US's NSA (National Security Agency) are not allowed to hack the hackers (i.e. attack those that attack you) but it goes on. Quite often, when an attack virus or Trojan is detected on say, a military network, it's allowed to sit on a secured system and monitored to see what it does. This gives the security agencies a clue as to where it came from and what it's purpose in life is. Once understood, a virus is created and sent back to it's source in order to learn more about it's creators and or destroy the systems on which it resides.

The US and Israel were very concerned about Iran's nuclear development. Now this is what I call a Trojan: The NSA and Israel worked together to produce a virus that would cause havoc with a certain piece of equipment used in Iran's nuclear manufacturing process. In short the virus was to cause the centrifuges used to separate substances, to run out of control and causing massive damage. The virus, called Stuxnet, was somehow allowed to get into the Iranian power plants network. Probably done by an innocent engineer downloading a so called "Update" (made to look like a normal expected Microsoft update let's say) for his laptop. This jumped to a pen drive which at some point was connected to a PC inside the secure network. It worked and the desired damage done. A minor problem however, that has come to light, is that the virus has infected machines elsewhere outside of Iran and has to be hacked itself to be stopped doing further damage.

The whole thing is actually quite worrying. We live on a fragile edge with the hackers and counter hackers doing their best behind us. Just be careful what you write in your next email is what I say. And, if I disappear soon after this is printed, I have said too much.

Chapter 16 Back and forward in time, chatting about some characters I think were great!

Sir Bernard Lovell's Jodrell Bank

I admire engineers that paved the way for so many things and yet seem to be somewhat unsung. Bernard was one such chap.

As a child and teenager, in addition to his love of physics, he had a passion for cricket, playing the piano, and was an organist at his local church. In fact, colleagues recount that during his working life, everything stopped for cricket, work coming second! Studying Physics at university, he obtained his Degree and then a PhD in '36.

He went to work as part of the Cosmic Ray research Team based at Manchester University. He was fascinated by Gamma-rays, X-rays, photons etc and the effects caused by Solar flares. This was quite ground breaking at the time and set the stage for investigating the beginning of the universe, Black Holes and so on. Ask Stephen Hawkin.

At the start of WW2, he was seconded to the Telecommunications Research establishment, TRE. He worked on the H2S radar system which was being designed to be fitted to aircraft. He worked closely with engineers such as Alan Blumein on the top secret and fundamental component of radar, the Magnetron. That's the bit that generates the high frequency energy, same thing that heats your food in your microwave oven. Lovell took over this work after the test aircraft along with several of his colleagues , including Blumlein, 'mysteriously' crashed in '42, killing all those on board. More on this under the 'Articles page' on my website. It's quite interesting…honest.

End of WW2, Lovell went back to his study of cosmic rays, at Manchester University. Finances were tight, and he managed to obtain ex-military radar, radio and other electronic gear, that he had been working with at the end of the war. The army provided him with an old arc lamp searchlight. It was supposed to be a loan, but they never saw it again. Lovell used this metal frame to build his first Yagi antenna array (type of aerial). The main problem he had, when searching into the night sky, was interference caused by local Manchester trams (things used to transport poor people) and ignition from cars. He persuaded the university to let him use some land near Goostrey, Cheshire. It was remote and distant from electrical noise. The area was called Jodrell Bank. Considering his setup was very Heath Robinson, working from the back of a van and trailer, the results were remarkable. This is where he set up his Observatory, which has become a household name. Using his cobbled together second hand kit, he managed to demonstrate how radar could detect meteors entering the earth's atmosphere. With government and university funding, he built what was at the time, the largest directable radio telescope. The "Lovell telescope" is still in use today and forms part of the MERLIN network. Multi-Element Radio Linked Interferometer Network. Interferometry, I hear you cry...what's that? I think it's the analysis of radio waves with a military, weapons bias. Too complex for me.

Jodrell Bank was used for, amongst other things in the '50s-60s, as part of the Cold wars' early warning system, keeping an eye on the naughty Ruskies. In fact, The Russians used Jodrell Bank to keep a track of their early-days space rocket launches. That always seemed a bit ironic, the fact they could stuff a rocket, complete with dog into space, but couldn't find them afterwards. Lovell claimed that during a working trip to the Centre for Deep Space Communications (Ukraine), during the Cold War, the Russians tried to bump him off by exposing him to a nasty dose of radiation. I am not sure if this was ever proved to be correct, but story has it that he wrote a detailed account of the episode only to be published after his death. Next time I speak to Putin, I will ask him and let you know.

Certainly as far as Britain is concerned, Bernard's contribution to Radio Astronomy is second to none. Many have built upon his early work and spectacular results of the telescopes sited at Jodrell Bank. He received his well deserved Knighthood in 1961.

Now this is praise indeed. Nigel Kneale, who wrote the brilliant Sci-Fi series, Quatermass, (Remember?), took Lovell's first name Bernard as the key player, Professor Bernard Quatermass.

Joe Meek (1929-67)...Recording engineer and record producer Brilliant chap!

For those of us that grew up in the '60s, Joe Meek may be a familiar name. I would call him one of the truly innovate record producers / sound engineers Britain had.

As a youngster, Joe realised his love for electronics, taking over his father's shed, modifying radio sets and making all sorts of electronic gear. After finishing National Service as a radar engineer, he worked for the Electricity Board in the Midlands. He had become very interested in music production and recording and managed to acquire a secondhand record disc cutting machine.

Joe left the Electric Board and joined a radio production company, making material for independent record labels and radio stations such as Luxemburg.

He gained much admiration following his work on the recordings of the Ivy Benson's all-girl dance band and Jazz Trumpeter Humphrey Littleton. Joe understood the importance of "engineering" the recordings. Remember, in the late '50s all we had were AM, Medium wave wireless sets and 78 rpm records. These had very limited frequency and dynamic ranges. So, to make the most of that, Meek compressed, topped-tailed the recordings, so that they sounded as good a possible. Humphrey Littleton recalled that on a recording of 'Band Penny Blues', Meek had compressed and uplifted the piano to give a big thumping presence. Littleton went mad, as recording was released without his hearing it first. He thought it didn't sound original. However, it was a smash hit!

In 1960, Meek set up Triumph Records and eventually moved into a 3 storey flat above a leather shop in London's Holloway Road. One room was a control room, with a lot of home-made recording equipment. 2 track tape machines, mixers, echo chambers, compressors etc. Other rooms contained various microphones and special effects gear. He would have piano and guitars in one room, drums and bass in others and some vocalists in his bathroom. Musicians often complained of getting electric shocks from microphones and guitars! Artists recording there included Lonnie Donegan (Cumberland gap), Johnny Leyton (Just Like Eddie), The Honeycombs (Have I the right) and of course The Tornados with Telstar, which was a number one hit in '62 and set Joe up. A classic example of Meek's playing around with recordings, can be heard at the beginning to the record Telstar. The strange space-like mechanical sound introduction, was actually a recording of an old tractor played at a slower speed and backwards. His recordings were over-dubbed many times, adding different instruments and vocal and effects. This detracted from HiFi quality. The record companies Joe used to press the records always complained about the quality of the recordings. But Meek knew this would produce a distinctive sound. It did. And, people loved it.

Joe was known to be hard to work with and during the mid '60s, was running into financial problems. He was convinced that Decca Records was bugging his studio to steal his recording secrets and would not let anyone into the building if he was not there. He had a real "thing" about communicating with the dead and would leave tape machines running all night in the hope of recording voices from the other side. He produced the Mike Berry recording "Tribute to Buddy Holly" following Buddy's death. Of course, being a homosexual in the 60s was a bit of a problem and in '63 was fined for importuning. He never recovered from this and it fed an on-going depression. Many musicians passed through his hands and It's said that Brian Epstein asked Meek what he thought of a group called the Beatles. Joe said don't sign them up. He agreed to record a small band as long as the signer was replaced. The signer being a young Rod Stewart.

In '67, in his recording studio, Meek shot himself after killing his landlady.

It is very sad thinking about someone who paved the way for modern recording techniques ended up like this. He was an early experimenter with compression, reverb, direct connection of instruments into recording desks, multi-tracking and so on. Phil Spector, well known for his "Wall-Of-Sound" image, said that he owed much of this to Meek's own inventiveness. I could go on and on, but not enough pages.

So, off to pop my copy of Telstar on the radiogram. Hope you are listening Joe.

Nikola Tesla, Brilliant engineer and inventor of the *"Death Ray"*?

Born in Croatia, 1856, Tesla was to become a massive influence in the area of electrical engineering.

After school and university, Tesla went to work for the Budapest Telephone company, where he helped redesign a lot of the transmission equipment, making amplifiers and repeaters which gave improved audio over the phone lines. Although he had over 250 patents to his name, this work at the telephone exchange was never registered, as he liked to keep his ideas stored in his head rather than on paper.

AC Power He moved to the USA and worked for the big electrical company *Westinghouse*. He went about redesigning their DC (Direct Current) generators to Alternating Current machines. This was to have one of the biggest effects on the way we live today. Using alternating current, power can be easily sent for miles. His work on high voltage equipment such as transformers and alternating generators (as we have today) set the path for modern power stations and the National Grid.

Radio He worked heavily on radio and, although we all think of Marconi as the inventor of radio, in fact, a lot of Marconi's work was taken from Tesla's patents and tested in court. Tesla won. In 1898 Tesla gave a demonstration of the first radio controlled boat! Onlookers were sceptical and had the boat removed from the water to see if there was a monkey inside following commands.

X-Ray After reading of Rontgen's discovery of X-Rays, Tesla went about designing an X-Ray tube to generate a powerful stream of rays. His work with high voltage transformers helped no end. Tesla noted that when playing around with his new high powered toys, there were some draw backs. He found that prolonged exposure of rays when imaging the body for example, caused skin to burn and damage to internal bodily organs. He also noted that sometimes when standing near to his X-ray tube, he felt a stinging sensation. This started him thinking of his next idea!

Death Ray This is the one I like. Again, falling back on his research on high voltage coils, Tesla came up with a design for the first real "Death ray". He devised that high energy coils could propel a minute stream of charged Tungsten particles over 200 miles, bringing down enemy aircraft etc. He tried to sell the idea to the UK and USSR. The theory was good, but with immense power needed to power a hypothetical machine and the lack of written down plans, it never happened, so it's said. Some people thought it was war profiteering and a stunt to raise his profile. Nevertheless, it lead, supposedly, to his labs and rooms being raided by the US Military.

Tesla died in 1943 at the age of 86, alone in the New Yorker Hotel where he had lived for several years. He preferred living in hotels as opposed to owning houses. Slightly an unsung hero I think, but a genius.

The Death Ray idea reminded me of the company in the UK that is developing a high energy radio transmitter whose purpose in life will be to stop vehicles from a distance. Great fun. You see, the cops would point the antenna at the fleeing car and at the push of a button, the radio beam would scramble the car's engine management computer, causing it to stop. It does actually work. A few refinements are needed. At the moment, it's the size of a phone box, so awkward for the traffic cops to carry around. As I understand, it's not Pacemaker friendly. If you have one fitted and you get in the way of the beam, you are turned into, for a short period of time, the Incredible Hulk.

There is a business opportunity here. Cars of the '60s and '70s never had engines controlled by computers. So, buy up the remaining stocks of these cars and sell them to criminal gangs who will then be able to escape unstopped by the Ray. I'm in the market for Cortinas and Mk 2 Jags by the way.

Alan Blumleine 1903-1942

Now here's a clever chap that not many people would have heard about. Amongst other things, his work was instrumental in bringing us stereo sound at the movies, stereo record recording, 405 line TV and Airborne Radar in WW2.

Cinema stereo sound

In his early career, Alan worked on the characteristics of the human ear and how sensitive it was to different frequencies and sound levels. This helped him design special circuits to improve speech over early telephone cables. One day at the cinema, he noted how annoying it was that the voices from the actors on film, did not follow their position on screen. Normally only one speaker was set behind the projector screens in the '30s. He developed a system called Binaural sound (later to be known as stereo). We worked out that recording two channels of sound using two microphones set on an axis of 45 degrees gave a realistic "spatial" sound reproduction. He helped design a system where this signal was recorded on the edge of the film and hey presto, the actor's voices would follow the images from left to right!

Stereo record recording

Whilst working at the Columbia Gramophone Company (later to become EMI), he started work on an alternative method of cutting sound on a master record. Up until the early '30s, if you wanted to cut a record, you had to pay big royalties to the U.S. company Western Electric. They had a patented design for the bit that converted sound to the cutting device used on a record. Blumlein designed a clever variation on this, using a moving coil system. No more royalties were needed and the sound quality was much better. He then modified this cutting head (using 2 coils working at right angles) which allowed him to cut a record grove containing stereo sound. The beauty of this design was that the record could also be played on a mono system with no loss of detail. Stereo records were born.

Television

There was a race to create a high quality "standard" for TV transmission and reception. Blumlein's boss at EMI, Shoenberg, set him to work full time on this.

Alan designed and patented many circuits to help and improve picture quality. Things like flyback scanning, black level clamps etc… all became common place in TV sets. Some of these principles are still used today. Blumlein and his team were instrumental in the 405 line standard that was adopted in the UK and many other countries prior to the war. It was the Full HD equivalent of today's telly. The 405 line standard ran for 50 years, being switched off in 1985. I remember the day well. I sat and watched the signal vanish on a 1959 Ferguson 506T. TV. Sad day!

There were numerous other patents filed by Alan and his team. Lack of space prevents me to rambling on. One patent related to the "Ultra linear amplifier" deign. He worked out that a simple connection at the output transformer of a valve amplifier, feeding an inverted sample of the sound back into the valve stage, created an amplified sound with extended frequency range and stability. This technique has been adopted ever since.

HS2 radar

During the early war years, Blumlein worked on the Top Secret HS2 airborne radar system. This was to increase accuracy in target bombing. Working with the TRE (telecommunication Research Establishment, Secret Squirrel stuff), he and several other senior engineers from EMI installed the sophisticated radar gear in a Halifax bomber and set off on trials from RAF Defford, 7th June 1942. After an hour or so, the Halifax developed an engine fire and crashed in Herefordshire, killing all on board. His work was continued by colleagues and was a recognised factor in shortening WW2. PM Churchill ordered a cover up of the death of the crew, even to family. Rumours of German sabotage were always maintained by Blumlein's wife.

So, there we have it. A short but brilliant life, gave a massive contribution. Thank you Mr. Alan Blumlein.

Hedley Jones..... Know about him?

As you know, I like writing about engineers and personalities that, in my opinion never received the recognition that they deserved. This chap did, sort of receive acknowledgements in his home country, but nowhere else. I am sure that 99.345r% of you won´t know who Hedley Jones was. Read this and you will.

I guess I write about this person because he helped set out the standard by which music of the West Indies became known for. I grew up in South London and had the privilege of bumping into someone who would become a very close friend, Martin Dennis, his parents being from the West Indies. I would say that Martin is probably the most intelligent, humane person I know. As time went on and I grew up, and realised what life must have been like for the first generation of West Indians coming to the "Father Land". That is a subject of another book for me to write one day. I won't go into this now, but the musical legacy; Mento, Ska, etc.... is a treasure. Only a few readers will appreciate why I keep my Blaupunkt Radiogram in tip top order. Oh, before I go on, I will add that I have only ever been in a car that was stopped for speeding once in my life. This was with Martin, at the wheel of his Mother's Rover 2000, NFG742D, automatic. I guess strictly speaking, given that we were both under age, we should not have been in that situation. The other fact I would add…. he was driving in reverse. The Police Man in is panda car was, I think quite impressed at Martin's car control. He got off.

Now then, Hedley, born in Jamaica 1917. He became very interested in music in his early teens and was a very practical hands-on kid. Times were hard and he made his own musical instruments, including a cello and banjo, teaching himself to play. In 1935 he moved to Jamaica's capital, Kingston. His work was very varied. Bus conductor, radio repairer, wood worker, sewing machine engineer. A local Mr. fix-it. Know how he felt. His main income was from being a proof-reader for a local newspaper.

As a lover of music, and inspired by the likes of Charley Christian, he played his banjo in a Hawaiian jazz band. He had heard that some open bodied guitars had been fitted with pickups for amplification, but not being able to afford such a luxury, he went about making what was to be probably the first pickup for a sold-bodied guitar, which of course he also made himself. He experimented with some coils from an old radio's coupling transformer that he had repaired, seeing that these could be fitted under the strings of the guitar. He used magnets from old telephone ear pieces to energise the coils. After many modifications, he settled on a design and started to make these for prominent Jamaican musicians of that era. The sound was great. This was probably about the same time as Gibson and Les Paul were doing the same thing in the USA!

Hedley knew that that in order to develop, he needed to educate himself in electronics. So, in 1943, middle of WW2, he volunteered and signed up as a Radar Engineer, where he ended up in the Royal Technical College Glasgow and did active service. After the war, returning to Jamaica, he opened a record store, Bop City, where he become popular for importing US jazz records, the likes of Luis Jordan and so on. He also repaired radios and electronic equipment.

He saw there was a market for amplifiers for producing good quality sound at dances, venues and so forth. With the knowledge he had gained in his National Services, we set about making, what was then to be some serious powerful kit. He realised that separating the treble and bass amplification was the way forward, feeding this into separate speaker systems. At that time, all you had was the Tannoy type horn speakers for general public address, no good for music. He experimented with beefy parallel-push-pull valve amps and set a standard that was to be used for years after.

The owner of a big store and dance party organiser "Tom the Great Sebastian" heard about Hedley's amplifiers and commissioned him to build a big sound system. It was great success. On the back of this, Hedley built systems for the likes of the fearsome gun-carrying Duke Reid, who was a record producer and owner of the renowned Trajan record label and sound system. He also built systems for a rival sound system "Sir Coxone", owned by Clement Dodd who later commissioned Hedley to build equipment for his famous STUDIO ONE recording studios. Collectively, everyone he worked with and for, said that he was the most influential person in the Jamaica's music and sound system history, spreading way beyond his homeland island.

Other areas of his expertise that were touched on were the likes of his design for the first traffic light system to be used in Kingston. He worked as an instructor at the Kingston Technical College and become involved in astronomy, designing lens and receiving awards for his work. He moved to Montego Bay in the mid '60s and continued with his love of music, being a band leader in high class tourist resorts. He became the president of the Jamaican Federations of Musicians which he held for some ten years. He continued writing for news papers and received awards from the Jamaican government such as Order of Distinction in Music. He sadly passed away in September 2017, a month before is 100th birthday.

So we have a diminutive chap from a tiny island in the Caribbean who helped set the stage for the famous sound systems and recording studios in the West Indies and elsewhere, that gave us lots of good music; Mento, Calypso, Ska, Reggae. And, probably made the first solid body guitar with an electric pickup. Salute to you.

Chapter Seventeen A bit aboutgGuitar amps

Marshall Amplification "Play it Loud, Play it Proud"

A nice aspect of my job, is working on and fixing valve amplifiers. And, there seems to be a steady increase in Guitar amps coming through the door, especially from the Spanish. **Marshall** is a name synonymous with such gear, so I thought I'd give you a little focus on that.

Gentleman Jim Marshall was in fact a good drummer. In the early '60s, he had a little shop in Hanwell, London, selling drum kits and all the stuff that goes with it. He was also into electronics. It goes that the likes of budding guitarists Pete Townsend and Ritchie Blackmore used to hang out there and badgered Jim into designing amplifiers that were to be "Loud".

Two brands of amps dominated; British made VOX and U.S. owned Fender. Both lovely amps, but at the time, expensive. Over a burger in a Wimpy bar one evening, Jim, along with an ex EMI engineer and a couple of others, worked out that he could make an amp to retail far less than the VOX and Fenders. The team used the Fender "Baseman" amp as a benchmark, but using European equivalent valves. ECC83 valves were used in the pre amp stages which had higher gain and would give the "Marshall Crunch" distortion sound when driven hard, that guitarists liked. The design also increased the treble sound which made it quite different from the warmth of the Fender. 6 prototypes were made. Unlike the copied Fender, the Marshall, was a stand alone amp in a box, with a separate speaker cabinet. This used 4, efficient 12" speakers made by Celestion and the cabinet had a closed back panel. This helped with providing the wanted punchy sound. Townsend, Blackmore and Big Jim Sullivan were amongst the first to own some of the 20 pre-production amps. The amp was named JTM45. Jim, Terry (son), Marshall, 45 watts output.

Jim was determined to drive costs down and entered into relationships with transformer makers, Drake and Dagnell. Up until now, the output valves were the American 6L6Gs. Jim changed to the British KT66, which had similar specifications and were interchangeable, but sounded harder. A young Eric Clapton used to sit in Jim's shop playing this new amp and loved the sound it made. He asked Jim to make a variation with a Tremolo feature and small enough to fit in his car. Jim did what he was told and the classic "Bluesbreaker" model was created. Clapton said it was this amp that gave him the great guitar sound heard on the '66 Beano album.

Jim listened to the up and coming musicians. Pete Townsend and John Entwhistle (The Who) were early adopters of the Marshall gear. They wanted more and more volume. Jim and his engineers set about designing a 100 watt amp. Basically, 4 output valves were used instead of 2, wired in parallel push-pull configuration. Bigger mains and output transformers fitted. This became known as the Super Marshall Lead model. Always wanting to go one better, Townsend asked Jim to build bigger speaker cabinets, which of course he did. These were massive and housed 8 x 12" Celestion speakers and packed a punch. The design was eventually dropped in favour of the classic 4 x 12" cabinet, mainly due to transport hassles. The big amps came with "Y" splitter leads so that amps could be connected together in order to give the power and volume required. This created the famous "Marshall Stack" image that has become the trademark of Rock bands. The bigger the stack, the better the band. Well…. Marshall did supply empty cabinet options for those bands that wanted to portray an image above their standing!

Build costs were rising and Jim started using different valves, components and manufacturing techniques. The KT66 output valves were replaced with Mulllard EL34s, giving a slightly different sound.

Jimmy Hendrix walked into Jim's shop in '66 and spent hours playing guitar through various amp combinations. Expecting that Hendrix would be asking to have the amps for nothing, Jim was astonished to hear that Hendrix would buy a healthy quantity of amps at full retail, on the basis that his road crew would be trained to maintain the gear on tour. Jim shook his hand off. Well, not literally, as that would have somewhat upset his guitar playing.

Many different models have been added since the '60s. No more hand wiring, made in the Far East etc... But, the brand goes on and still designed in the UK. Next time you are head banging, look up and see if Jim is looking down smiling at you.

Wonderful world of guitar amplifiers

The nicer side of my working life is repairing vintage guitar amplifiers, for two reasons I guess. One, I like repairing equipment for people that appreciate a good job done on their treasured piece of kit. You see, us musicians are a funny bunch. We get attached to our instruments. The other, I get to pay with hot sweaty valve equipment, a first love for me.

We all have favourites, mine is **Fender**. Two reasons; Sound is normally quite "pure" and build quality. However, cheap production costs in PRC leaves it's mark. Earlier amplifiers were wired point to point. i.e. no PCBs, components being connected via tag strips and suspended on the valve bases themselves. This made serving easier and components were generally rated in excess of what was to be required of them. Many of the early amplifiers from the 60s, used a fiber board with studs which were used to mount components and cabling. The problem here was, that if the amplifier was exposed to moisture, the fiber board was anhydrous (absorbed moisture) which resulted in all hell breaking loose. High voltages are used in valve equipment and this caused tracking within the fiber board. If left unchecked, the tracking would end up leaving carbon deposits, which then made the amplifier almost useless. The cost of this sort of repair is very expensive. Older amps have a very much "Fender" sound. It is quite true and normally lacking in purposeful distortion, making them very clean sounding for accurate guitar work. Lovely amps included Fender twin reverb 135, Fender Vibro king, Fender tone master

Jet city

Have seen quite a few of these newish amps. On the surface, they look a bit "budget", but lift the bonnet and it's quite a different thing. They are well made, use good quality components, transformers and no sparing on resistor wattage where required. The PCBs are thick and tracks very substantial which is a must for valve gear. Well worth the money.

Peavey

Not a great lover of these amps. Seem to be over complicated for what they do and many models are not built with servicing in mind. Quite often, the rating of components is right on the edge of working conditions. Typical example would be bridge rectifies used in valve heater circuits. They run hot, go short circuit and unsolder themselves. I always replace with much larger chassis mount devices. Some models such as 5150, VK100 etc... have additional valves in the pre amp and drive stages. This creates noise amplification, and with microphony (vibration sensitive) causes a right racket if the amp is sat on a speaker cabinet. If I am in a good mood, I sometimes make small modifications to introduce negative feedback at high frequencies to reduce the effect.

Marshall

Good old Jim Marshall. We all know about Marshall amps and the thing that sold these in the early 60s was the way they could be driven to give a distorted sound. The later models are sometimes hard to work on, but generally are good. Sometimes the manufacturing build quality lets them down with component use and fitting not being appropriate for valve gear. I Quite often you look at the circuit design and layout and think "this designer learned his stuff at college, not taught by his Granddad with years of experience".

Sinmarc

I have a soft spot for these Spanish amplifiers. These were made in Barcelona during the 60s-70s. They are well made and used the best components available at the time. A few of these models, 50 and 100 watt, are quite blatant copies of Fender models. They use big transformers and especially the output transformers have plenty of overhead giving good damping and bass response. Some models even had a "Magic Eye" to tell you how loud you are playing. Ideal for deaf guitarists (most of them).

Laney

No a great fan. Build quality, PCBs, resistors etc all could be more substantial on a lot of models. I often find that the copper tracks on the PCBs are not very tolerant to being resoldered and reworked. Laney as a company (Headstock Distribution) are good to deal with and helpful.

Vox

As long as its an old original amp, then these are lovely. The equipment was well made and thought through, great design. Vox was sold out and most of the equipment is manufactured in the PRC. An iconic brand that featured heavily with all the British rock-n-roll bands in the 60s and 70s. Now the gear is like most of the budget kit you buy. It's a shame.

Chapter Eighteen Some companies that always interested me

The company that gave us more than just the Public Address system!

I've written about various engineers who have left their mark on the electronics industry. So, here's another one that shouldn't be forgotten – Guy Fountain.

Now this chap started a company, around the late '20s, later to be called called Tannoy. That word would become a colloquialism for public address systems, "Can Hugh Jampton please come to reception". Bit like calling a pen a Biro and vacuum cleaner a Hoover.

The original name was Tulsemere manufacturing company. Fountain designed a new type of rectifier (converting AC mains to direct current), to enable households to charge the big lead-acid batteries used in radios of that time. The device used dissimilar metals, Tantalum and an alloy of lead. It was a massive success and he set up the company Tannoy in 1932, an abbreviation of the metals used.

In the '30s, development on loudspeakers was moving fast. Tannoy was producing speakers that could handle a considerable amount of power for the day. A range of microphones had also been developed which were ruggedised, with special audio characteristics. These were used by the Army and Navy etc. The distinctive Tannoy logo was always seen on speakers installed around all around Butlins, Pontins holiday camps. The company also made all the clever audio switching equipment and valve amplifiers. The components used were always "top end". The systems were very reliable. I remember, in '79 just missing a chance to buy all the old gear from the Orchid ballroom in Purley way, Croydon. It had done 25 years service and was going strong.

In the '40s, Chief engineer Ron Rackham, very bright chap, saw that there was a growing interest amongst home enthusiasts, requiring better high fidelity sound equipment. He wanted to get away from the usual speaker cabinet with a bass unit and a separate higher frequency driver (tweeter), mounted elsewhere in the box. His thinking was that if the sound was originally recorded from one location, it should be reproduced from one point. He set about designing a speaker whose Tweeter was part of the centre assembly of the bass unit. His dream of the full spectrum of sound, bass to treble coming from the same position was realised. Tannoy called this arrangement "Dual Concentric Cone", as the voice coils and cones of both bass and tweeter speakers fitted concentrically together. The speaker, named

"Monitor Black", big 15" diameter, was launched at the 1947 London Radio Show and was very well received. The basic design has gone through changes, Monitor Red, Gold, and latterly Gold Reference. These 50 year old speakers, if looked after, still sound fantastic and don't need huge amounts of power. In those days valve amplifiers produced 10-30 watts a channel. A set of these speakers will set you back a couple grand easily!

Tannoy still dominate the high-end of the HiFi market and the speakers are generally first choice when it comes to professional high quality sound systems. London Palladium, Sydney Oprah House to name but a few, all the way down to the home enthusiast wanting to create as faithful music reproduction as possible. Recently, an interesting award winning system called Qflex was announced. This uses their speakers driven in a special way as to "steer" the sound to specific areas, being used in stadia and music venues etc.

Guy Fountain retired from the business in '74. Tannoy still has a research and development centre at it's HQ just outside Glasgow. The company is now owned by the TC group of Swedish companies. This group of companies specialise in various avenues of audio recording, processing and reproduction. They all pool their resources enabling the design of quality equipment aimed at home use all the way up to massive venues and military applications. Tannoy is an important part of that group.

Next time you hear an airport announcement, it might be over a Tannoy!

Thorn Electronic Industries

If like me you're the wrong end of 50, you may recall many household electrical goodies showing the brand Ferguson, HMV, Marconi, Ultra etc.

It all started with Joules Thorn who was a salesman for the Austrian Oslo company, selling gas mantles. Those little furry things that vanished when you poked them. The company, like the product, went pop and a chap called Alfred Deutsch, who made lightbulbs, persuaded Joules to start a lightbulb business. The Electric Lamp Service Company was born in 1928.

After a while, they acquired a company, Atlas lamps. Atlas made bulbs for projectors and so on. By 1936, the firm had expanded massively and were probably the biggest manufacturer of light bulbs, fittings and associated parts in Europe. Some of the lighting equipment was beautiful and reflected the Art Nouveau styles at the time. By now the company was renamed Thorn Lighting.

Having cash in the bank, Thorn started to diversify and acquired Ferguson Radio Corporation. This was a US-Canadian company making well built rather fancy looking (Big flash US style) radios in the UK. Ferguson sets of the '30s were relatively expensive, mainly due to the "Valve tax" the government placed on each radio at the time. This was calculated on the number of valves a set had. Being US based, they stuffed upwards of 8 or more in a set. Valve makers; Mullard, GEC, Osram etc, started developing valves that had 2-3 separate assemblies in one glass bulb, doing the job of 3 valves. Clever way round the tax ah?

Ferguson became a household name for TVs, budget radios, big radiograms and so on. Sets were normally quite fancy and appealed to those living on the East side of big towns and cities. In '60 Thorn bought Ultra Radio & Television. This company had been making TV and radio sets since before the war and had spread out into all kinds of other electronic equipment.

By now, Thorn group was probably responsible for 40% of the radio and TV sales in the UK. Thorn TV sets were reliable and rental companies favoured them as a result. In the late 60s, Thorn had developed the first UK fully transistorised colour TV, the 2000 chassis. This set the benchmark for other set makers such as GEC, Bush etc. who were still using rather unreliable valve designs that were partial to the odd internal bonfire. The 2000 chassis was very complex and I recall fixing a few of these. The service manual suggested an engineer having a treble Teachers before taking the back off. The set gave a good picture and behaved itself.

Thorn acquired the well know Radio Rentals company. This company at it's peak, had upwards of 500 shops around the UK, had in excess of 2 million customers and thousands of well trained engineers and support staff. With the availability of colour TV transmissions in the late '60s the demand for TVs was massive. Great outlet for Thorn's TV manufacturing. Others companies snapped up were the well known DER and Rumbelows group.

In the '70s, less that 2% of households had a video recorder. Radio Rentals saw an opportunity to add this product to their portfolio. They settled on a VHS machine made by JVC. Thorn branded it under the Baird name, a company also purchased years previously. It was a solid machine and reliable and certainly helped VHS standard become the more dominant system, with pre recorded tapes being made widely available. Although, other systems such as Betamax and Philips 2000, were technically superior.

Thorn went on buying companies such as Kidde (fire alarms), Mazda (lightbulbs), Tricity (electric cooker company) and the HMV record group and even Kenwood (food mixers). Thorn went through many changes and as a result of the increase in reliability of TV sets and VCRs, purchase of equipment was more attractive than renting. As a result, the rental parts of the business were sold off to smaller household chains and holding companies. There we go. A huge company started by a gas mantle door to door salesman.

Here's a food mixer joke; I went to a kitchen store and it took ages to get served. I shouted "can someone

sell me a food mixer??" A voice from behind the counter said "Kenwood?" I shouted "yes he will do"....
I'm wasted in this job.

BSR *The automatic record player!*

Dr. Daniel McDonlad '05-"91.

If, like me, you are a '50s-'60s ite, you probably remember stacking records up on an automatic turntable, watching the first drop, being played then another dropping and so on. Now, various companies made auto-changers (as the Yanks called them), but in my book, **BSR** (Birmingham Sound Reproducers) made the best, at an affordable price and could be repaired and well thought through. Unlike some of the competition, these machines were easily maintained and worked reliably. Garrard players were well made and then, Collaro... These never worked even when they left the factory. They were responsible for many a humble engineer turning to hard drugs.

The founder of BSR, Dr. Daniel McDonald was a clever individual. He started out his engineering life working for BTH, British Thomas Huston (Big Thick and Heavy as it should have been known). He saw a market for quality mains and audio output transformers, for use in radio and Public Address gear. In '32, he was making these from a small workshop in Blackheath, in the Black Country (that funny Birmingham area). As his transformers we so good, he decided to design and sell his own PA (public address) systems which normally included a record deck for the music element. His first machine was the **Ampligram**. A fine looking highly polished case, housing a 25 watt amplifier (massive for those days), a microphone and record deck. The adverts were lovely, *"designed for halls where 350 couples can dance the night away"*!

Daniel always was insistent that, as far as possible, everything should be made in house. He increased his factory to make motors for the record decks, cabinets etc.

Between '47 and 58, **BSR** became the biggest supplier of turntables. These ranged from single play decks to the famous "Monarch" range which could play a mix of different size records, stacked in any order. The deck used a clever little "finger" which, as the record dropped from the centre spindle was knocked down and detected the diameter of the disc. This set the pickup arm to start playing at the right place. Brilliant to watch in action! Tens of thousands of record players and radiogram manufacturers used this range of decks. Another key item designed was a light-weight pickup (UA8). This was based around a small quartz crystal, which when vibrated by the stylus in the groove, produced enough power to drive only one valve in the record player. This drove down costs. Prior to this, pickups had big magnets and weighed the same as a family car, with shopping and quickly wore out records. He also established a sales force to target the U.S. market. There were some battles with Zenith and other companies but product was sold and stood its ground.

Due to massive growth of the business, factories in Northern Ireland were set up and elsewhere in the UK, this had its problems. Daniel was a very hands-on manager and put a lot of store in keeping his staff happy. It was almost a "lifestyle" company. Dances, outings, events were always being held and staff loved working there.

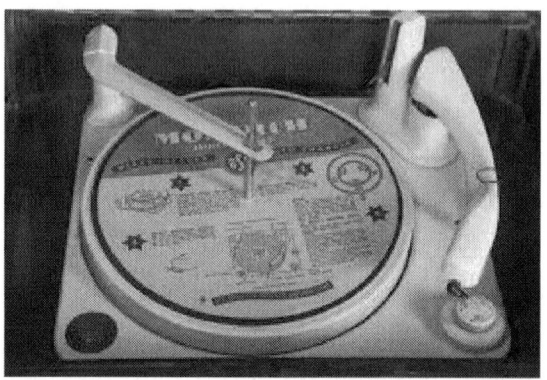

The product range was huge. "Music while you work" systems, office intercoms, Police car loudhailers, to name but only a real few. In some ways this was to be the beginning of the downfall.

A great little tape deck was also part of the portfolio. The TD1,2,4 etc were simple, reliable machines and at a good budget, were bought by every tape recorder manufacturer. This meant the Joe-Public could own their own tape machine to record their best tunes from

Two Way family Favourites on a Sunday after dinner, or record Uncle Jack singing down the boozer. Again, to keep things in house, McDonald bought the company who made the tape heads.

By the late '60s – '70s, there had been lots of changes. McDonald had taken a back step and new Whipper-Snapper management had not done a good job of things. New products like 8 track decks were being made and clever electronic record decks that made the coffee as well, were being churned out. With a company jet and other extravagancies, one could see what was going to happen.

By the '80s things were not good. Competition from the Far East etc. finally closed the doors on what was one of those great British companies. So there we go. If you grew up playing your Lonnie Donnigan records on your Dansette, please, say a quick thank you to Dr. Daniel McDonald.

Chapter Nineteen   Old and new

82 years of TV – Now in Colour! An update

Well, it is 80+ years since the first official TV broadcasts were made from Alexandra Palace. Logie Baird and EMI battled it out during an evening's viewing in the late '30s, seeing which TV system would give the best results. Baird's system was based on a frightening mechanical, 240 line system. If the main part, a spinning disc broke loose, it would have taken out half of Southern England. EMI's version used all electronic cameras that could be mobile and the resolution was 405 lines. EMI won. This system remained in use in the UK until 1985. Just shows you how good it was. Of course, this was a black and white system, but colour was on its way.

Colour TV had been around in the USA since the late '40s. RCA had developed a system for transmitting colour signals working with the NTSC, National

Television Standards Committee. This standard was also known as *Never Twice the Same Colour*, (I am being sarcastic, unusual). The system had flaws. The early colour decoder circuits in the TV, struggled to resolve the correct colours. People's faces would slowly turn green and blue over the course off an evening's viewing. Sets had a colour "hue" control to compensate, meaning chubby Chuck would have to get up from his god-damned-son-of-a-bitch chair to adjust the set every hour or so.

France adopted a colour system called SECAM, *Séquentiel couleur à mémoire.* It was the first European system to be developed, really only used in France. Like most things French, including the women and *Citroën cars*, it was bloody complex and a nightmare to work with. Smoke a packet of **Gauloises,** down a bottle of **Dubonet***,* was the best way to attempt a repair to these complicated circuits.

Britain and most of Europe used a colour system called PAL, Phase Alternating Line. Developed in Germany, it was a simpler system and worked well, being able to display existing black and white transmissions.

1967 The start of UK Colour TV!

BBC and ITV had been gearing up for colour, on the back of the new UHF 625 line TV system. Radio Times proudly highlighted programs that were to be *In Colour.* This was a massive boost to the TV rental businesses. Early Colour sets were not reliable. "Dodgy Bob" round the corner, had been bodging black and white sets up to now. Colour sets demanded expensive tools and some proper knowledge. You had to have a good aerial and a strong signal. If not colour would fade in and out.

Philips G6

In '67, Philips released a monster of a TV chassis. It was a mix of valves and transistors. Very hard to work on. I would rather change an engine in a tug boat single handed than work on this set. Valves that gave off X-Rays, sterilising the engineers' nether regions. Took 4 of you to get the set upstairs.

GEC

Late '60s-'70s GEC brought out a range of colour sets. Again, Hybrid, a mix of valves and transistors. I have a soft spot for these sets. I bought them, ex-rental, by the car full, change a few parts that were known to go faulty and away the set went for another 5 years. They did however, have a tendency to catch fire, which some people found annoying.

Thorn 2000

Here we go. '68, Thorn (Ferguson, Ultra, HMV..) designed an all transistor colour set! It had 14 different circuit boards, massive cable looms, loads of preset controls to twiddle in the back. The TVs were good and set the way for most makers' development plans. Other Thorn chassis, 3000, 9500 ran into the late '80s and were all good.

Mullard, Mazda made most of the colour TV tubes. Depending on how much the set was used, you might get 5-8 years before the picture went soft, out of focus, as the electron guns lost their emission. Lots of little companies set up, selling "re-gunned" tubes. You took your old tube in, the vacuum within the glass tube would be released carefully, the glass cut and the old gun assembly removed. The new gun would be glass-welded to the tube and a vacuum pump used to remove the air. Back to the workshop, fit the tube into the set and spend the next day trying to set it up so the picture was watchable. Some of the re-gunning firms were rubbish! I made sure all the sets I bought for refurbishment and resale had Toshiba tubes fitted. Shame, but these lasted so much longer.

We now have 60"+, high definition, 4K 8K, 3-D, Cinema sound and can watch TV on the move. I wonder what Logie Baird is saying about this?

The TV Remote Control

Well, I have written about many interesting developments and engineers relating to the electronics industry, but this article is about the history of the most important development since man invented the wheel. The TV Remote Control. Small hand held device that empowers its' owner, controlling the lives of many.

Remote controls in fact, have been around for years. Nikola Tesla had put forward a design for sending radio wave to control a distant machine. There is a patent for this, going back to 1893! However, as radio and TV became more popular from the '30s onward, not having to get from your chair became important. Company called Philco came up with a battery operated low power radio transmitter that would change channels on one of their wireless sets. It was, however, huge, heavy, unreliable and expensive. Apart from that it was good.

A U.S. company, Zenith claim to have had the first TV remote. Well, it was remote from the TV, but connected to it by a thick cable. This was a draw back as everyone tripped over it. The control enabled you to turn the set on and off, change volume and channels. It did this by controlling several motors in the TV connected to the channel and volume knobs. It sounded a bit like a printing press when operated. The name for this device? "Lazy Bones". In fact, Dynatron in the UK was supplying a similar device for their up-market sets.

Zenith made further developments and of course coming up with great names for them. "Flash Matic" in '55 saw a range of TV sets that had a true detached remote facility. The TV had 4 photo-cells in each corner of the screen. You had a torch with a narrow beam, that when pointed and flashed at the corners, changed the station, volume and on-off. It was better than the first attempt, but you had to have a good aim and it didn't work if the sun was coming in the window. In '56, Zenith came up with the first Ultrasonic remote, the "Space Commander". This box housed a mechanical contraption with several keys. When these were pressed, the mechanism plucked different length pieces of metal inside. This resulted in a high pitched "ping". This is why the remote became known as a Pinger. The TV had circuits that would detect the high pitched ping. The note generated of each key was different and the TV used this to change channel, volume etc. I suppose it was better than the first two attempts but it too had flaws. The high pitched ping would annoy the pet dog who in turn would whine and this would be seen by the TV as a valid signal and change channel. Dropping keys on the floor and outside wind chimes also played havoc with the evenings viewing!

The Ultrasonic remote was developed over the years and remained the mode of operation until the mid '70s. Remotes now had ultrasonic transducers, driven by circuits that generated audio tones higher than humans can hear. The tones were grouped in different groups of pulses, allowing the TV to decode many different commands, for Teletext pages, picture settings etc. TV manufacturers got together and agreed on a format of these signaling codes. In the early days, it was a free for all and different remotes would interfere with different makes of TV. War would range in households that were lucky enough of have TVs in adjacent rooms!

By the mid '70s, we now had the infrared remote. This, instead of an audio transmitter, was a LED (light emitting diode) which creates a beam of light which our eyes can't see. This beam is detected by the TV's infrared receiver. The light from the LED is pulse modulated, governed by a small pre programmed processor circuit in the remote handset. Again, these pulses are coded by the frequency that they are sent and by manufacturer type. There is a huge range of allotted codes, as you could have all kinds of remote controlled equipment in the living room. Each device has to be able to receive and decode its' own specific set of commands. The code structure is not as simple as you think. For example, you can keep the volume button pressed and the volume will continue to increase. Other functions will need to be a one press only function, like the Mute, one press on one press off, for example. All clever stuff.

Let's have black & white 405 line television back!

In the 30s John Logie Baird was playing with his version of television. This was an electro-mechanical system and consisted of a metal disc with holes punched in it spinning in front a neon tube. The resulting picture was 3″ square, orange and viewed through a magnifying glass. You had to connect this contraption to your radio, tuned into the shortwave TV transmission. The resolution was 30 lines! Ok if you were watching Come Bloody Dine With Me.
The BBC and EMI decided that this was not good enough and eventually settled for 320 or 405 line system, all electronic. 405 lines won and that standard lasted from 1939 'till 1985! Logie Baird never recovered from this.

Up until '57, you only had BBC, then came ITV. The war stopped TV production and transmission. German bombers would use the powerful signal from transmitters at Alexandra Palace as a beacon to bomb London. After the war, everything started again and the Coronation was responsible for a huge sale in TVs and that set the scene. Early sets had a 5"- 9" round tube. If you wanted a bigger picture, you hung an oil filed magnifying glass in front of the set. It gave a slightly bigger screen, but you had to sit directly in front!

In the 50s, better and larger tubes were available. 12 and 17" was common. Watching TV was fun. Turn the sets on about 10 mins before your programme (6-5 Special!) to allow the set to warm up and settle down. Sets in these days bristled with controls that would need a tweak throughout the evening. Vertical hold to stop the picture rolling up and down, horizontal hold (line hold) to stop the picture shooting from one side of the screen to the other. Height, Width, linearity and so on. Brightness and contrast of course. All of these needed playing with from time to time depending on how warm your house was and what the electric supply was like. In the winter, it was common to have 20% less mains. This caused your picture to shrink, roll. It was fun! The more expensive sets had a "white spot suppressor" control. This was supposed to get rid of interference caused by trams and cars with dodgy ignition. Did it work? Nope.

Now, if the set went wrong, you would have all heard the "repair man" say, with a sharp intake of breath "it's ya picture valve me duck". In fact, there was no such thing as a picture valve. Picture tube (screen), yes. The so called "picture valve" was normally the little EHT rectifier, which gave the high voltage needed for the screen. In early sets, this was a little undernourished valve, EY51 by name, which sat quietly on top of the line output transformer. It had a hard life and expired regularly. Changing this chap took about 15 mins. and it cost 10 shillings plus tax. The grubby repair man in his Ford Pop' van, would keep the set for a week, scratch the highly polished cabinet and charge you £11.2s.6d for a new tube (screen).

'64 saw the arrival of UHF 625 line transmission. The sets were more expensive, you had to change your aerial and could only get a picture if in the right area. Then, '67 and we had colour! Sets would cost £350 (a family car was £500ish). You had to be a proper engineer to fix these. The EHT section (nasty high voltage 25,000 volts) was contained in a metal screened tower housing big beefy valves, PY500, PL509,PD500, EY501. The latter two gave of X rays when working. You didn't stand in front of this lot when working on the set with the metal safety screen off for obvious reasons and it was waist high and would make your gentleman's area sterile! These sets were prone to the odd bonfire. Philips, PYE and GEC were great for this but all fun. Sets in the 70s used a mix of valve and transistors. They were starting to get more reliable and that killed off the rental shops as people were now happier to invest in something that lasted a year or so before going pop. I refurbished many of these sets in the late 70s and early 80s. Replace all the known nasty bits and the set went on for another 5 years. You could always tell what a household had for Sunday lunch. The high voltages attracted dust, smoke and grease. A set on the bench would smell of roast lamb and Embassy no.5. A combination of smells I will never forget!

Chapter Twenty A bit about radios

When Portable Radios were not as portable!!

Ever since Marconi got fed up with painting his bathroom and decided to invent radio, we have wanted to carry one around with us. In the 1920-30s, sets were nice and big, using valves and components that one would find in a normal table top set, indeed, the same. Two small differences really; one being that some bright spark had screwed a handle on the top and the other, you needed batteries. Now, to operate the valves, batteries had to be substantial. You required an accumulator cell (big jam jar, like a small lead acid car battery), giving a couple of volts to drive the valve filaments (to make them hot), a high voltage HT battery around 90-120 volts and a smaller battery for Grid Bias. Add the lot to a big square frame aerial and a speaker. Chuck this in the back of your Austin 7 or Rover 12, which could just about manage the weight (women had to ask their husbands to lift the wireless in and out the car because in those days they weren't very strong you understand) and drive off to your picnic. On a good day, you could find a station to listen to, BBC's 2LO (coming from the Strand in London) or the Home service. This lasted for about an hour and a half before the batteries went flat. When you wanted these recharging, you would take them down to your local bicycle shop, where they would have the charging equipment to do such a job. They were set up for charging motorcycle batteries and so charging radio accumulators was a good money spinner.

In the '50s Mullard created a new range of small 7 pin valves "D" series, which were designed to use little power. Now we had a radio that *was* portable. Still had to have batteries, 1.5V dry cell for the filaments and a 90V one for the HT. These batteries cost in today's money about 20€. Who remembers the nice bright blue and red oblong batteries?! Now you could get three or four days use from them and a bit more if you put the batteries in the oven for 20 minutes with the Sunday roast, when they became flat. Most of the sets had a "mains" option. So you could give up the silly potable idea and leave the set indoors. A common design was a mini brief case style with a lid that opened up. As the lid opened, this normally brought the set into life. The lid also housed an aerial.

The battery company EveReady made lots of these sets, along with a round weatherproof set made from saucepans, painted in blue Hammerite paint. These were aimed at the African market, were insect, little finger and spear proof.

By the late '50s and '60s, transistors were used in portables. Sets were now tiny about the size a large pack of fags, with a little speaker that squawked away, and came with a fake leather carry case. They used a PP3 9V battery which was cheap. Who remembers seeing on the back "Empire Made" and "Made in Hong Kong"? When GB was GB.

A great British company called Roberts made lovely sets. Wooden case, big speaker, tone controls etc. They sounded like the big wireless in the living room. In the '60s their chief engineer, Mr. Hacker left Roberts to start his own radio company and called it Hacker, funnily enough. No surprise that the resulting sets were solid and perhaps over engineered. Sounded great and worked well. As time went on however, competing with the Far East closed the business. Shame.

By the '70s and '80s, most of the UK radio manufacturing had been hived off to China and elsewhere, with just some of the brand names continuing, having been bought by holding companies. Bush, Alba, Murphy and so on. But never fear, we had the Ghetto Blaster, for those with a chip on one shoulder to help keep the balance whilst carrying it on the other.

Classic radios of the past

Wireless set from the '40s-60s, were shackled to a degree performance wise, given that transmitted material was limited to Medium and Long Wave. This gave you an audio bandwidth of about 8kHz which by today's standards doesn't give you much to listen to. To make the most from this limitation, set makers designed clever tuner, amplifier sections decent efficient speakers and of course great cabinet designs. Having several old radios in the shop that tend to attract attention, I thought I would show some classic examples.

♦BUSH

DAC90A

Now this little set was very popular between the late '40s and '50s. It had several variations. It was a budget set with a Bakelite case (early form of injected moulding plastic). It was a simply constructed set and was, as was common in those days, AC/DC. This doesn't have anything to do with the sets split sexual preference, but meant that it could work on any of the different mains supplies that were found in various parts of the country, before the national grid standardised supply. When serviced, these sets work very well and will easily outlive their owner.

The "Round" Ekco

Mr. E. K.Coles company of Southend-on-Sea (sarfend as the locals would say) Produced lovely sets. To keep costs of case making down, the company invested massively in high pressure Bakelite moulding plant. This enabled them to make cabinets any shape as required. They came up with a range of round shaped sets, ranging from budget types to electronically quite advanced. Some had round dials with hidden illuminated tuning pointers! Anyone who recalls "Till death us do part", will remember seeing one of these sets on the sideboard behind Alf Garnet.

Beau Decca

In the '40s Decca made some spectacular radiograms, namely the Beau Decca and Decca Deccola. The wooden cabinets housed several loudspeakers who's sound was channelled around the room via angled flutes. Amplifier sections used big PX4 and PX25 valves and coupled with a revolutionary HiFi pick, the sound quality beat all that was around at the time.

"Reverbeo"

Always producing good sets, there were some interesting features. In the late '30s-40s and range of sets came out known as "Mono-Knobs". A mechanical nightmare of cables and pulleys gave the user a single control that when rotated, tuned the set in, when moved side to side changed the tone and moved up and down changed the volume. All well and good 'til something went wrong. The Philips service manual had a detailed section explaining how the repair man could hang himself halfway through trying to repair one of these sets. By the '60s and with the wide coverage of FM/VHF transmission, some really flash sets had been brought out. These had split tuning systems so that you could pre-select your medium wave and VHF stations, stereo amplifiers , twin speakers. If you gave up the family holiday and chopped in your Mk 1 Cortina for a Ford Pop', you could afford a stereo decoder so that you could listen to the Third Programme In full stereo! For those that didn't want to do that, there was a variation on one of these sets called "Reverbeo" The set had a spring reverb unit fitted between the stereo speakers, giving a large concert hall sound effect. Actually sounded great. Dolby eat your heart out. Sorry, Mr. Dolby passed away recently.

McMurdo

This was a radio manufacturing company in the U.S. The sets were sold in the U.K. made under license, normally by companies such as Ferguson. The sets were over the top, with no expense spared. The chassis was normally chrome plated and loads of valves used. Unlike sets made in the UK in the '30s-50s, the U.S. sets were not subject to the British Valve tax. Purchase tax on the radio was based on the number of valves used. This is why manufactures such as Mullard we good at making one valve that actually did the job of 2 or 3. The McMurdo Silver 15-17 is a typical example of a brute of a set. 15 valves, BFO for short wave etc. Push Pull output stage packing a big noise into a 122 energised speaker. I was lucky to come across one of these sets. It is in se today. In 1937, the price tag was 137 Guineas. A huge amount of cash.

McMurdo Silver 15-17

Blaupunkt Cocktail cabinet Radiogram

Blaupunkt (German for Blue Spot) was, is still I guess a high quality radio manufacturer. In the mid to late '60s, it was the in thing to have a radiogram. Sort of music system of the day. These sets had quite a complex circuit, squeezing every bit of quality and volume from the usual line up of valves used in that period. These grams had a wonderful warm bass response and sounded great. They became very popular within the West Indian communities and would have been pumping out Ska and Calypso at house parties in the late 50s. My one, model Arkansas 59 is used daily.

Chapter twenty one

Radio during the war years!

With the onset of WW2, industry in Britain changed dramatically. Typical affected areas were of course Radio and TV. Many of the manufacturing and technology companies who had been designing and making radio and TV sets, now had their efforts firmly trained on developing and building equipment for the Army, RAF and security services. Companies such as Ekco, Bush, Murphy and Cossor shelved all their domestic activities in readiness for making radio sets for Lancaster and Hurricane bombers etc...

Prior to '39, TV had developed quickly. BBC and EMI decided to settle on the "high definition 405 line" TV standard. Transmissions had started from sites such as Alexander Palace in North London. Albeit TVs were still basic, had small screens and had not sold in volumes forecasted. The TV transmissions were halted as soon as war was declared, as the strong signals from the transmitters could easily be used as direction finding beams by those naughty German bombers. The companies that had been developing TVs were naturally well placed, with their knowledge of making TV cathode ray tubes (screens), to build radar sets.

All this left the government with a problem. It was imperative for the public to be kept informed about the war effort, propaganda it could be said. But, a massive short fall of wireless sets, only 70,000 having been made in '41, with an unrealistic 250,000 scheduled. 10% of sets owned were broken, waiting parts. This made things a little awkward. Component manufacturers, making coils, valves, capacitors etc were working flat out to address military requirements.

In '43, Churchill and the BBC met and it was decided that a specification for a "basic" radio should be drawn up ASAP. The result was a set designed by Murphy Radio's Chief Engineer Dr. Reynolds, using clever techniques to reduce component count. It had one band, Medium Wave, and a dial marked with BBC stations. The design was such that the chassis could use valves from different manufacturers, even from the U.S., as production and availability in Britain was sporadic to say the least. Several manufacturers were licensed to make the sets, but they all looked the same. A manufacturer's code on the back was the only way of knowing. Sets sold for about 12 quid. The "Wartime Utility Radio" was born. Under 200,000 were sold. Germany had a similar counterpart called the German People's Receiver, Volsempfanger.

In order to keep up the supply of radios for the public, it was decided to import basic sets from the U.S. About 5,000 were imported, keeping back 20% for spares. The main problem was that sets from U.S. ran on 110 volt mains. In Britain, the voltage was 220 volts. This meant you had to "loose" 110 volts somewhere, or suffer a bonfire. This was often achieved by the use of a resistive mains cable. The cable from the set to the plug wasted the unwanted 110 volt which generated heat, about 40-60 watts. On a cold day, the cat could be found sitting on the coiled up cable behind the wireless. All ok until it had a pee. Moggy, exit stage left.

Following the end of the war in '45, set makers soon re-tooled to start making radios from where they left off in '39. Most makers completed the mothballed sets they had, which were sold at a lesser price. The following couple of years saw some interesting designs coming through, with manufacturers using the new technology gained, having been making military equipment. Some companies such as Cossor, Plessey, set up subsidiaries divisions, making equipment for the military, on the back of their new learned knowledge. By now the government had massive stock of electronic equipment, receivers, components and valves alike. Desperate for cash, it was deemed that this should be sold off. Stores sprang up selling Army surplus. Lisle Street in West London was a great place to buy all sorts. For a while, TV manufactures were worried about sales. You could buy all the parts you needed from these surplus stores. Ex radar display CRT screens, EF50 valves (developed for military VHF radio) and so on, enabling you to make you own TV. Many Blue Prints were around showing how to do this. Only draw back was the picture displayed was via a green 6" screen! As a kid, I bought all my bits from a surplus store in Kingston. It was an Aladdin's cave. I bought a Russian test meter, still use it today. It always reminds me of my Russian ex girlfriend. Heavily constructed, not particularly attractive, but performed well.

Chapter Twenty two

HF radio Communication - Anyone out there??

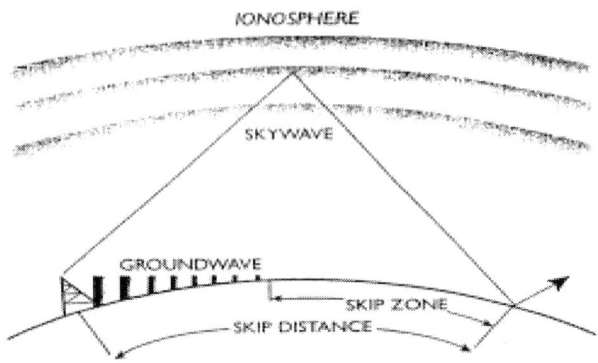

So, HF radio (High Frequency), what's that all about? Well, we all have heard about amateur radio and Morse code etc. HF radio, was and still is a good way to make contact with people across continents. Let's forget about the Internet for a moment. The radio frequency band we are talking about is around the 1.5 to 30 Megacycles. I say High frequency but by today's standards it is quite low. Our terrestrial TV gets to us using frequencies in the 800 Megacycle range.

The beauty of this HF range, is that radio signals radiated from an antenna connected to a transmitter, behave in an unusual way. The earth is surrounded by a band of charge particles called the Ionosphere, like a big balloon with the Earth in the middle. This invisible band sits at around 80-400Km above the Earth's surface. This distance varies depending on time of day, year and temperature and so on. Now, with higher frequencies such as used in FM radio and TV etc. these signals tend to travel in a straight line from the antenna. Given the curvature of the Earth, you could have a line of sight from the antenna which is about 30 miles. So, trying to receive the signal further than that, you are a bit stuffed. However, the frequencies in the HF band have a tendency to travel upward and hit this Ionosphere. When they do, some of the energy bounces back to the Earth's surface and again can get reflected back up to the Ionosphere and so it goes on. How exciting I hear you all cry. The interesting bit is that the angle by which these signals are bounced back and forth depends on the frequency of the radio wave. The higher the frequency, the more acute the angle. Now then, go and get your school geometry books. Imagine a triangle forming points, one from the ground antenna, one from the Ionosphere and one back to the Earth's surface where you have your receiver. As the angle is dependant on frequency, the position where the signal lands on the Earth's surface to be received changes. Lets say you are transmitting In Albox on a frequency of 8 Megacycles. Someone in the U.K may be able to hear you. Jolly good. Change your frequency to let's say 4 Megacycles and the U.K listener goes deaf, but someone in Sweden hears you loud and clear. The other factor is as previously mentioned, the height of the Ionosphere changes with time of day and so on. This therefore, (trigonometry basics!) changes the location where the signal from Albox can be heard. You could get to Iceland let's say (continent, not the store).

This signal can also get caught within various strata of the Ionosphere and bounced around the Earth. This is where radio amateurs are able to talk to each other, from France, USA, Japan and so on. I was going to add Birmingham, but who would want to? The good thing is, that you do not need loads of power to achieve this. 100 watts or less of power could get you all over the planet.

Before the days of Internet, satellite communication etc. HF radio was the way to communicate over vast distances. During the War, HF was used to send encrypted Morse code, telemetry and speech, keeping in touch with all concerned. The colonies relied on radio to keep in touch with the mother land and the BBC World Service had a network of transmitters set up for this reason. Companies such as Marconi had sophisticated mathematical models to work out precise frequencies that could be used in order that, let's say the West Indies could hear the Queen's Christmas broadcast. An interesting example for the use of HF radio was a thing called Weather Fax. The Met Office used to (still do) transmit a facsimile signal containing low speed facsimile picture of weather fronts. Anyone with a radio receiver and Weather fax machine could have a print out of what was happening area by area. This was used heavily by shipping fleets and so on.

Nowadays, HF radio is more used by amateurs who like the buzz of finding intercontinental mates to chat with and utility companies who need to keep an electronic eye on their systems which are spread out over vast distances. Even Morse code is no longer being used.

The Tug Boat

Whilst working for Muirhead in the late '70s to mid '80s, I had some good fun with HF radio. I had to engineer facilities to allow interfacing facsimile machines to HF radio equipment. In the middle of this connection were encryption systems, commercial or government. As anyone could "listen in" on the broadcasted transmissions, they had to be secure. Any facsimile signal that was picked up could only be decoded if you had the right encryption equipment and associated codes etc. Using HF radio, allowed these facsimile transmissions to be sent where telephone networks were not in place. Some of the projects were with oil companies. Sending lots of drilling information via satellite links in those days was very expensive, HF radio was cheap. The oil company, Shell, was investigating a potential drilling location way off the Irish coast. We had to test to see if the radio frequencies provided by the authorities would allow encrypted facsimile transmissions to and from the location at sea and the offices in Shannon, on the coast.

The only way of achieving this test, prior to the big drilling ship being sent, was to set up a station at sea (on a tug boat would you believe) where the exploration was to take place. Along with this, a receiving station at the offices in Shannon. I put a team together, a couple of engineers from my company and a couple from Marconi, who was supplying the radio gear. Like an idiot, I said I would do the tug boat end.

I recall it being a damp miserable day. The chap from Marconi and I had lugged all our gear on board the boat moored at Kinsale harbor. We set off along the river Shannon out to sea. At that point, my experience of boats was limited to my father's inflatable with a 10 H.P. Volvo Penta outboard. We set all the equipment up in a spare cabin and managed to get things working on a limited power supply of 115V AC from the tug's generator. The Marconi man busied himself fixing a huge antenna on various parts of the boat's superstructure. At that point, the Skipper of the boat asked if I had experience at sea. Sure I said, thinking of the fun with dad's inflatable. Why did he ask that I wondered?

As we left Shannon and edged into the Irish Sea, the ride slowly went from a gentle roll from side to side to something less attractive. I listened to the ship's radio playing the weather forecast. With every sea area mentioned, getting nearer to where we would be doing the tests, the gale force warning got worse, ending up as a gale-force 8 imminent. By this time I had started to sweat, in a most disgusting way, I was dizzy, felt sick. The skipper laughed and said the best thing to do was to lie flat on my back on the bunk in the cabin. This gave me relief. The chap from Marconi also had a similar reaction to the sea sickness. We shared the cabin and I am not sure what was worse; the smell of his foul feet when he took off his shoes or the fumes from the ship's engine. I found that per hour, I had a window of working ability of 40 minutes. 20 minutes to lie down flat and recover.

The tests lasted for a couple of days which seemed like a lifetime of torture. Luckily things seemed to work. We headed back to land. However, whilst dismantling my equipment, the ship heaved heavily and I dropped a spanner in one of the ship's fuse boxes. There was a big bang, the radio went quiet and the skipper started shouting as to why the Decca Navigator system was dead. I was not flavour of the day I can tell you. It delayed our journey home which for me, was a disaster.

We eventually got back to land and remember getting off that bastard tug boat, kissing the tarmac on the drive out of the dock. Two problems remained. One, I could not walk. I stumbled, fell over and could not run up stairs like I had always done. "Sea Legs" I was told. Took me a week to feel steady again. The other problem, I had to tell the big Irish chap who we had rented a house from in Shannon, about a little accident we had. The house was used as a base to set up a temporary receiving station. The guy from Marconi, unbeknown to me had erected a massive antenna, bolted to the side of this small house. The storm that had ruined my nice sea trip had hit land and a gust of wind took the antenna down along with half the top of the gable wall of this nice chap's house. He didn't seem too bothered. I was expecting a smack in the chops as starter. But he laughed and said it would be okay, just an insurance job. Don't you just love the Irish?

Now for the big boat

After proving the encrypted facsimile transmissions worked, Shell in conjunction with the South Eastern Drilling Company, sent a big drilling ship (the SEDCO 472) off to the Irish Sea. Off we went again to set up a station on land and on the ship. I did the ship end, on the thinking that big ships don't bob up and down like silly tug boats. Apart from nerve wracking helicopter journeys to and from the ship and sharing cabins with American oil workers who where 7 foot tall and the same wide, the time spent was quite nice. I always slept with one eye open just in case, well you know....

I set our equipment up in the big radio room on board. There were lots of nice toys to play with. There were two radio operators, taking turns on watch. Both huge guys, smoking big cigars and calling everyone Chuck, Buddy etc.... I got on okay with them.

Now, I have only been physically thrown out of two places in my life. One; the Empire Ballroom Leicester Square in London, along with the bass player from the band I played in. We had a tiny drop to drink and set the fire alarm off, shutting all the music down. The other was the radio Room of the Drilling Ship. One early morning, as usual, I couldn't sleep. So, as I had done several times in the past, I had got up and used the ships radio to call my fellow engineers back on shore just for a chat and to see if we had pulled down any more houses. This particular night I sat myself in the radio room which was unmanned. Not unusual for that time of early morning. I powered up this big transmitter. ITT-STC kit, about 3000 Watts if I recall. I put out a call to land and waited for a reply. All of a sudden, I was aware that I was flying through the air, eventually landing on a gangway outside the radio room. One of the radio operators pulled me up, pinned me to the wall and proceeded to explain in a very gentle calm manner, what could have happened as a result of my transmission. You see, I didn't realise what was going on. During drilling a test hole earlier that night, they came across hard rock. When this happens, what they can do is to drop radio controlled dynamite charges down the drill hole and set them off by radio control. I was then shown a long line of radio controlled dynamite charges, all lined up along the deck, ready for use. As these, as I said were radio controlled, the last thing you want is any spurious radio transmission going on, especially from the ship's 3000 watt transmitter. RADIO SILENCE and been ordered. I didn't know did I? I guess it could have been a very big bang and a lot of insurance paperwork to fill out. Luckily I am here to tell the tale!

I'm too young to die

Whilst on the H.F. and facsimile radio bit, it reminds me of a time working with the C.I.D and Special Branch in Wales. It was at the time in the mid '80s when the I.R.A was posing a big threat. We supplied encrypted facsimile systems to communicate with the ferries crossing the Irish Sea. Passenger and car details could be sent back and forth safely from ship to shore using encrypted facsimile.

I had to set up a station on the top of a hill in Wales where good radio communications could be achieved to the ships sailing across the channel. As this was an area that was known to be monitored by the I.R.A. I was assigned a plain clothes guy to keep an eye on me. I got to this radio shack on the top of this hill and the chap from the C.I.D. gave me a hand held walkie-talkie and said if I thought there was a problem call him up and he would come running. Sod this I thought, it's a bit above my pay grade. But, all fun.

Anyway, got to work and about 5 hours later all was up and running. Then, the metal door to cabin shut with a thump and I couldn't open it. I heard something outside. With no windows to look out and a door that I could not open, I decided to call P.C. Jack Regan. 15 minutes or so went by and I heard the sound of a 4x4 of some kind. The C.I.D. guy opened the door and laughed, calling me Wanker. I looked outside and saw a herd a cows, one of which, being inquisitive had walked up to my door, shut it and sat in front. How was I to know??

However, trust me, with the tension twix the US and North Korea, our internet and satellite communications could be wiped out. The best way to communicate in that event..... HF radio.

Chapter Twenty three

The National Grid – 90th Birthday. Electric at our fingertips

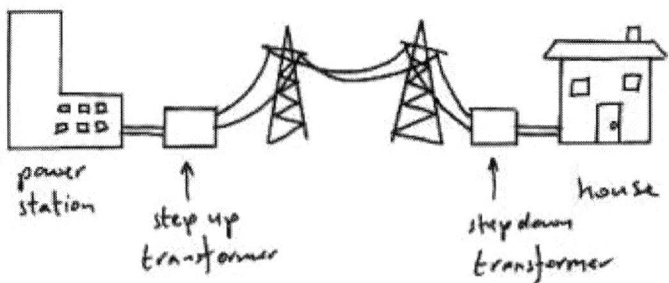

A simple thing as plugging the telly into the mains socket is a given to most of us. But, what goes on behind getting the 230V AC power to the socket? I hear you all shout. Quite a lot as it happens and it was not so simple years ago.

In the late 1800's, U.S based Nikola Tesla had designed a system to generate and distribute 3 phase Alternating Current (AC). Even by that time, the demand for electric light, motor driven equipment was increasing rapidly. Edison also had put plans forward for a power distribution system, but using Direct Current (DC, like out of a battery). Big battle, but a long story short, Tesla's AC system was by far the best and most logical. Here's why.

DC is OK, but once you generate it, you are limited to how far you can send it. It has to be sent by cables and at the voltage you generate it at. Cables possess a thing called resistance. If you send, let's say 200 volts DC over 2 miles. What you get at the end may be 170 volts depending on the current (Amps) you draw. Overall power is equal to Volts x Amps. What a waste. So, you had to have small individual generating stations near all conurbations.

With AC (Alternating Current), you can now use things called transformers. This makes sending power over great distances much easier. The wasted power that is suffered over the power cables is a function of the current being pulled through them. The overall power we want at the distant end is equal to Volts x Amps as mentioned earlier. So, if we whack up the volts and use less current, the loss in the cables is far less of a worry for the same amount of power.

Britain in the early 1900s, saw lots of small generating companies, all independent, with all sorts of power ranging from 100 – 200 Volts, both DC and AC. A nightmare when buying lamps, motor driven equipment and so on. Those of us of a certain age will remember seeing settings on the back of the TV / Wireless saying 200, 220, 240 V, AC-DC (nothing to do with the set's sexual leaning). In mid 1920s, the government commissioned a Scotsman Lord Wier, to solve the problem. He put forward a network, based on Tesla's system. The UK would be supplied with 50Hz AC power, distributed by overhead cables with a voltage of 132,000 Volts. He called it the Gridiron. The "CEB" Central Electricity Board was born. By

1933, over 4000 miles over overhead cable had been deployed, linking the best power stations, about 120, together.

As said, the beauty of AC (the power swings negative and positive 50 times a second, which is why your guitar lead buzzes when you plug it in), is it can be transformed. From the power station to a major town, the overhead pylons carried 132,000 Volts. When this got close to a small town, it would be transformed to 11,000 V. Nearer to houses, it would be again transformed to 230V and fed to underground cables to the home.

This Grid crisscrossing the country allows any power station to hook onto it and feed in energy. Now, the AC voltage is swinging back and forth 50 times a second (frequency, called Hertz) and before the power station connects, the generators (Alternators) have to be fully synchronised to the voltage on the grid. This frequency is linked to the speed of rotation of the generators. Get this wrong... run and hide. Generators could rip themselves to bits, with lots of paper work to complete before going home.

The frequency accuracy, 50Hz, is important, although not so much these days. In the past, many items used electric induction motors. The speed of these is determined by the frequency of the voltage. A clock for example, would run slow if the frequency went down to 48 Hz. An accepted tolerance is 0.1% and that's varied, + and - over 24 hours making an average.

The monitoring of power demand of the grid is crucial. Companies study the TV schedules and social events, in readiness to produce extra power when needed. A high audience TV show for example ends at 21.00 hours. Kettles, microwaves are put on and demand can go up by 10%. Therefore power stations have to be ready the boilers, open the gas taps, pull out reactor rods and chuck some more pallets on quickly.

The Grids 90th birthday! Big cake, cooked in an electric oven, of course.

Chapter Twenty four

Potted history of the mobile phone!

Earlier, I wrote about how the public telephone system started, with Mr. Strowger's invention, the automatic telephone exchange. When I wrote about this in one of the magazines, my email inbox burst with excitement. I thought it may be a good following on to talk about the dawn of the mobile phone.

Mobile telephones have been with us since the late 1940s. Bell Labs in the US, started work on a system which would integrate a hand-held army type Walkie-Talkie into the public telephone network. These were not quite like the SmartPhones we have today. Big batteries, valves, 5 foot long antenna and weighing half a ton, you needed to be wealthy and strong to own one. You would have to call up the local radio operator and ask nicely, to be put through to the desired land line number. The controller would basically patch the radio-transmitter equipment to a switchboard. The conversation was only one-way at a time, known as switched simplex. You would operate the "Press-to-talk" button and release it after shouting "Over" to hear the other end. After a few minutes, your batteries would go flat, your arm would drop off and your bank account would be empty.

In the '60s-'70s, if you wanted a telephone in your car, again you had to have deep pockets. In the UK, a trunked radio network had been established. Similar network as used by the Military. A number of radio repeater stations, linked together, giving very limited coverage in main towns and designated important areas. The radio telephone was quite a beast to be fitted into the car. The size of a couple of shoe boxes, a nice chunky handset and holder and a substantial rear wing mounted aerial, about 3 foot long with a loading coil the size of your fist half way up. Not car-wash friendly! Anyone clocking the car however, knew you were a grown up. The early versions had to be manually switched to the strongest radio site. You would call up the radio operator and asked to be connected. Again, the two parties could only talk one at a time. Push the button to talk and wait for a second or so for the transmitter part of the phone to get going, then say "over" when you wanted to hear the other end. Later generations of systems were more sophisticated, with two-way speech, know as duplex. Telephone numbers could now be associated to the radio phones, allowing making and receiving calls to public telephones, without operator intervention.

Now how about this. Arthur C. Clarke wrote a paper in 1959 setting out a society where people could talk to each other, anywhere on the planet by dialing a number on their personal hand held communicator, which would also be able to give you a global position!.

The Cellular system we now have, had been evolving since the '50s. A network of linked radio "cell" sites, all connected to the public telephone exchange network. The clever bit, is that a cell phone logs itself on to the strongest cell depending on where it is. The cell knows that particular phone is under its control. The network controlling system therefore knows where to route the incoming or outgoing call to and from. What happens when I drive out of my cell? (on hands-free of course), I hear you shout. Well as part of the management system, your phone and the cell are continuously looking for the strongest signal. If the phone sees a stronger cell, as you drive from one area to another, it ready's itself for a "Handoff". Without your knowledge, the phone transfers the call and connects itself to the most appropriate cell.

UK had its first cellular system in '83, Known as 1G. It was an analogue system using the classic "Dell Boy" brick handsets. Speech quality was not good and security was a problem. It used frequencies freed up from the old 405 line TV system. With a radio scanner, one could hack into a call and copy phone numbers to program another phone. The 2G digital system started in the '90s, using much higher frequencies and digital encoded speech, audio quality was better. We now have 3,4,5G data web and so on. Interesting fact that the SMS (Short Message Service) was an adjunct only to be used to advise a service engineer of his next service call. Who would have thought...?!

Chapter Twenty five

A salute to British HiFi !

In the 50s, Britain acquired quite a name in the HiFi world. The war effort had halted manufacturing of commercial radio and by the mid 50s things were starting to look a bit more rosy, with more money around to be spent.

Pre war, Medium Wave radio was the thing, record players were heavy clunky machines with pickups that weighed same as a sack of coal rather than grams, records didn't last long. Developments in stereo sound reproduction had been well underway in the '30s, helped along by engineers such as Alan Blumleine at EMI, who had been instrumental in radar development for the RAF. Decca came up with a full range record pickup. Lighter and capable of high quality reproduction.

In the '50s, crystal pickups were common place. They were cheap and allowed kids to afford their own record players (Dansett!). Three companies dominated record player manufacturing. BSR (Birmingham Sound Reproducers). The decks were simple budget machines but worked well and could be fixed with a coal chisel and some string. Garrard decks were solid, better quality and cost more. Then Collaro.... Now these decks were aimed at the high quality end of the market. Unfortunately the bloody things never worked consistently, even when brand new. Collaro was responsible for 89.5% of radio repairers being admitted to mental institutions as a result of fighting with these dam things.

Tape recorders were becoming popular. You could record your best tunes off Two way Family Favourites, save buying the records (although you heard the clock ticking in the background and granny doing the washing up, because you had the microphone next to the radio, facing the wrong way). You could also record yourself, down the Pub singing! BSR again came up with a budget deck, TD1, used by many tape recorder manufacturers. Apart from the little round spring that used to break, they were great machines. A quick call to BSRs' office, speak in Brummie and 3 replacement springs would be in the post. If you didn't mind spending a months' wages, you could buy a Truvox, Brenell or Ferrograph. These had an extra replay head, so now you could make an echo chamber, what fun was had! Collaro also made tape decks...say no more.

Post war amplifiers were becoming more sophisticated. Mullard were producing good valves to cater for this, EL84s and EL34s, still are the preferred valve today in the enthusiasts' ear. Makers such as Quad, Leak and Rogers were making first class gear. These makes command a high price today.

Loudspeakers were a crucial part of your set up. They were efficient given that you only had 10-20 watts of power to play with. Tannoy, Warfdale, Celestion and Lowther were excellent examples. With cleverly designed cabinets, every last drop of power was made use of. They sounded great. You would be amazed what you can do with just a few watts. You don't need buckets of power, honest.

FM radio started broadcasting around the mid 50s. You were free of interference and listening to concerts sounded like you were in the orchestra pit. Leak came up with a very stable tuner. It was based on a coaxial line resonator, called a Trough-Line. Other tuners drifted and you would chase your station around the dial every half hour but these were rock solid. You had a nice green "magic eye" valve to assist in tuning in to the station. It looked impressive and was a big help if you were stone deaf. Stereo radio was introduced the early 60s. And, with the addition of a decoder to your tuner or radio, you were flying. BBCs Third program started stereo programs first, followed by the Light program.

In the 50s-60s, there were many small back street companies, employing 10-20 people producing good equipment. The designs were normally based around Mullards' published circuits which were to promote the use of their valves. It was a good move. I suppose it's good testament to these companies that their products costing say £35 in the early 60s now sell for £1,000 plus, and that's after giving over 40 years of service. By the 70s, many of the names had given up competing with the imported Japanese equipment. The businesses were sold off and their names used to badge mass produced rubbish.

The best of British HiFi!

Post war Brits' were starting to have a bit more disposable cash and with HiFi audio recording, record player and tape recorder technology gathering momentum, one area for people to have fun was investing in "HiFi".

Two companies dominated the British manufacturing scene. QUAD and LEAK. If you were a proper grown up and loved music, your home system had to be based around one of these makes.

Quad

In 1936 Peter Walker set up Quad. Making amplifiers for sound reinforcement, public address systems etc. he had a good understanding of how to get a faithful reproduction of amplified sound. He used well designed transformers and circuit design which was not too common in those days, mainly cathode coupling of valves. Being a bit techie. The Quad 1 set the scene for quality domestic reproduction. Audio I mean. The amplifier design ran for some 15 years with an inconsiderate small hiccup in the middle called WW2. In 1953, with the Queens coronation etc. The Quad 2 was released. This was a cleaned up version of the old Quad 1 using nice big GEC made

KT66 output valves. KT was short for Kinkless Tetrode, which was a valve design to overcome an inherent distortion in power output valves. This valve dated back to the mid '30s and were originally used as radio transmitting valves. This amplifier was so jolly good, its production ran for some 18 years, with almost no change. In the mid '60 production gave way to the Quad 3 which was a transistor design.

In '57 Quad came up with a revolutionary loudspeaker called an "ESL". Electrostatic loudspeaker. This looked like a 2 foot by 3 foot curved room heater. It comprised of 2 thin layers of Mylar plastic films insulated from each other. A high voltage, 1,500V, was applied across them and the audio from the amplifier caused the films to vibrate in sympathy to the music. The resulting sound was a very accurate rendition of what was squirted in. Great for classical music, but a little lacking in bass.

Leak

Harold Leak established his company in '36, making amplifiers, similar to Quad, for the public address and cinema business. With a small team, he designed an amplifier called the "Point one". The amplifier could reproduce quality audio with as little as 0.1% distortion when thrashed at full tilt. Unheard of (no pun intended) in those days. Again, these amplifiers used the KT61 and KT66 output valves. Various designs followed that were equally as impressive. The TL12 became a "standard amplifier" to which all others were measured. The BBC had a variation specifically made and were used in studios all over the world as the preferred amplifier. Leak went on to produce stereo versions in the late '50s, the stereo 20 and 60. They had separate control units, giving the users flexibility in designing their home system. That was in the days where the consumer was credited with enough intelligence to put together a system that suited their ears. Leak's success in my view was the quality of the output transformer (you have to be proper engineer to design a good one), which connected the amplifier to the loudspeaker. These amplifiers were over engineered and stood the test of time. In fact, a brace of any of the above will sell for hundreds of pounds, normally being snapped up by the Japanese, who hold British sound equipment in high regard.

The leak Point one

Chapter Twenty six

How servicing has changed!

Talking about radio and TV over the past chapters, make you realise how things have changed from a servicing point of view.

In the 50s-60s, electrical goods were made, to a degree, with serviceability in mind. Radio and TV sets had nice plugable valves. A change or swap around was a quick job and given wages verses cost of goods was on your side, you could make a small living. One could even change a set of brushes, costing 2 shillings, in a vacuum cleaner motor and make a profit. Hands up who has tried to change them in a Dyson? Plastic and pretty but rubbish. I'd rather replace a bus engine single handed. Before coming to Spain, a friend asked me to look at his Service washing machine complete with mangle. It was purchased in 1956. In 2005, it needed new brushes! That was it.

House calls were quite straight forward. Everything you needed to fix 80% of TV faults could be fitted into your case. 10 of the most common valves, a hand full of dropper resistors and capacitors, meter and soldering iron and that was it. Remaining 15% could be fixed with some string and sellotape, whilst the last 5% were taken to the bench for surgery.

Quite often, things were not as they should be. Mr Twiddler was a favourite. You always knew when you found loose or missing screws on the back of the TV, the local "expert" had been there first. Valves would have been swapped around, all the preset controls played with. Even bits removed. There was no point in asking if anyone had looked at the set before. "Oh no me dear" was the reply. Up the bill by 50% was the answer.

Occasionally, there would be the odd pop and puff of smoke. Not a problem if you were working on the set on the bench, but in someone's house, a list of explanations to hand was useful. "oh yes, that sometimes happens with this model" and "oh, has it ever done that before?". I recall replacing a large capacitor incorrectly, the result being that after the set was on for a minute, there was a huge bang the offending item shot straight through the back of the TV and made a mess on the curtains. That was a bit awkward. I bought a job lot of aerosol switch cleaning fluid from Dodgy Bob. Good price I thought...wonder why? A lady complained that some of the station select buttons were sticky on her TV. Ah I thought, this calls for a spray with my new tin of magic squirt. After 10 seconds, I noticed the plastic front of the set melting. That was hard to explain away!

There were some light moments. I was called to a house where the radiogram was dead. It didn't take long to realise that the lady of the house was a "professional" shall we say. She asked if I wouldn't mind keeping quiet in the front room whilst she attended to one of her clients. It didn't take long to fix the set and on the turn table, by chance, was Honky Tonk Woman, Rolling Stones! Well, try and stop me playing it. I did of course, at full volume. 15 minutes later, the lady returned. She gave me a £2 tip, not from her but from her gentleman friend who found the whole experience much more gratifying knowing that someone was downstairs playing records! True story. Oh we did laugh.

Servicing nowadays is a little different. Certain goods are cheap and cheerful and not meant to be fixed. Washing machines, tumble dryers are assembled then welded together and painted. You try changing drum bearings, control units. It's a nightmare. Parts can be expensive, add a piddly amount for labour, it's not cost viable. Take a large LCD or plasma flat screen. You had spent 1000€ 3-4 years ago and you might get the average 3 years before you call me to have a look. Parts, depending on what needs to be done, could be a significant percentage of the cost price. People object to paying. We now have to be very resourceful as to how we carry out repairs. Always go down to component level where possible to keep cost down. Don't forget, LCD tends to be a little more reliable than plasma. Oh and by the way, insurance companies are wise to the fact that most plasmas fall off the wall after 3 years and don't pay out, if you get my drift!!

So that's why there are few 'little repair shops' in the high street. To make a living is hard work and you won't see proprietors of these shops driving Jags or playing with speedboats.

Chapter Twenty seven
Customers!

This is really a follow on from talking about how servicing has changed. We all need customers. After all, without them, there would be little income. I guess after 40 odd years of being in the service industry, in various guises, you get quick to judge someone as they come through the door. You have to weigh them up to get a feel for the hassle factor of what awaits you. You look at their body language, eye contact, clothing, what type of watch they wear, the car they drive, how they address you etc. etc.... Add all these things into a complex equation and that sets you up for the job in hand.

Customers can have a great way of saying things. It could be a passive aggressive way of saying that the last job you did was not very good. A typical example would be; "Hello mate, you looked at this TV a little while ago and it´s gone again". Now I am quick to pull this all to bits. This is what I generally find… The set came in 2 years ago, with a sound fault for example. The set brought in today has a fault with the HDMI socket on the back that has been broken away. So, nothing to do with the original fault and if it was, it is 2 years down the bloody line. Warranty or no warranty! Depending on what mood I am in, I sit the customer down, offer him or her a vintage Malt Scotch to sip whilst I draw an analogy along the lines of…. "If you took your car to the mechanic to your local well priced and professional mechanic to have the brakes sorted and 2 years later the rear wiper motor failed, would you say "Hello mate, you looked me car a little while ago and it´s gone again?" The answer generally is NO. After this little interchange, I set about beating the client across the head with an old heavy oscilloscope that I retain for this very purpose.

So, I have compiled a record of "What customers say". All that I have written here is *actual*, nothing fabricated. It even makes me laugh when I read it back to myself. It IS all true, that I can be sure.

I am not sure if the Spanish sun affects people, but I seem to get my fair share of odd balls in the office. Below are just a few examples:

The helpful
"Alo mate. Tv wont start, the red light flashes. Google says it's just the **capacitor***". Now, on the bench I have a small pot with old capacitors in it. Label on the*

pot says "Just capacitors". So I reach for the pot and give the customer one. "There you go sir, let me know if that works". Odd silence and look of puzzlement follows normally.

Same applies for resistors and of course. The "FUSE", Customers come in with a blown fuse and say "you got one of these?". So, I reach to the pot which is full of obviously blown fuses, nice and blackened. "there you go", I say.

"My set's dead, so I think it's the on-off switch". Cant help this one…. I have a box of 1950s Wylex porcelain and Bakelite light switches that are about 4 inches square. "That's a bit of luck, I have a replacement, it's a bit big but works well".

Mr. Overflow of information

Phone rings. "Yeah, got a Samsung with a common fault"…… Silence. … My last reply was, "oh, is that where the set dematerialises from one side of the room and reappears on the other during East Enders?" … Customer (really happened) " Ur no, aint done that".

The Spanish are very tight with the truth.

Chap comes in carrying an old Philips CRT set. I translate…"The TV has no sound". "Ok", I say, did the sound go off all of a sudden?" "oh yes" says owner. Later that day, off with the back, to find the speaker is missing and the audio output I.C. blown in two. In comes Pedro later and I show him the vacant space.

As usual, you are met with a shrug of the shoulders. I said to him that there are robbers going around breaking into peoples' houses, removing speakers from TVs. Spaniards do not get our humour. Typical found on a skip job.

There was the Yorkshire man who spent 3 minutes telling me his DVD player just needed "fettling". After the repair, he came in to collect it and read the invoice. "Carried out Fettling procedure in accordance with Panasonic tech bulletin 5". He told me he thought it was that and went off happy.

Customers coming in the Ipads, Tablets, phones etc. that won't do what they should. Now, I have a box of old valves on the desk, big chunky ones, KT88s – 5 inches high and 3 inches stout etc... They come in and ask what the problem was and I cannot resist showing them a valve, saying "it was the picture valve sir". With something 4 times the size of the tablet, the customer still says thanks and goes off happy.

"Picture keeps jumping" says the customer, as soon as the phone is answered, not even an introduction as to who is calling..."Obviously an issue with the vertical hold .." he helpfully added.
Answer ..'have you tried putting a heavy table lamp on top of the set' ? "No" he said, "Not yet".

The other morning, I had a chap come in, and, I even made myself laugh.

He came in and said "I have a Sharp TV in the car"
So I said without thinking, "be careful not to cut your
fingers on the edges then". Poor sod stood there
looking vacant.

Nice to have customers with a degree of humour. This
bloke had me going and he did a good job too. The
phone rang, "Hello, Zeta Services.." said I in my usual
helpful manner. The chap "said hello sir, you have
been shagging my wife" For once I was quiet, thinking
carefully. "Right" he said, "now I've got your attention,
I've got a problem with my laptop". Oh we had a
laugh.....

Hello, do you repair TV's?"
"Yes"
"well my aerial has fallen off the roof, can you come
and fix it?"

When I was very much younger, I was asked to go to
a house to fix a radiogram. Nice lady came to the
door and showed me in to the front room. Record
deck running slow, BSR, needed oiling. Fine. Then
knock at the door. She asked me if I wouldn't mind
keeping quiet whilst I was working n the gram as she
had a "customer". She would be done in 20 mins.
Being young, I didn't know quite what was meant, but
eventually put 2 and 2 together. Anyway, got the set

198

working and rummaged around in the 45 rpm stack of records and came across Honky Tonk Women. Well, couldn't resist putting it on. When the lady came in the room she smiled and called me naughty. She asked how much the repair was and she paid. Then she gave me a 5 pound note, saying it was a tip. I refused saying it was far too much. She then said it wasn't from her, but from her "customer" who had found things more exciting knowing there was someone listening!! True.

Was called to a private old people's rest home to look at a juke box. A nice 1956 Wurlitzer 2000. With my head in the back, trying to remove the amp and PSU, I was aware of sound of running water. I thought someone was pouring a cup of tea. I looked round and saw an elderly gent standing in front of my tool case weeing in it! I can't quite recall what I shouted but a nurse wandered up saying "oh so sorry, Mr. Johnson does get confused". I'll give me f...g confused I think I said. My Fluke Multi meter was never the same after. Again, true.

A Monday afternoon, get your own back.

Phone rings…
Customer: Right, bought my telly in 2 weeks ago and the sounds gone again.
Me: What's the name?
Customer. Sony
Me. No, your name please, or the invoice number.

Customer. Jeff. 302114, you tucked me up 80€
Me. Strange, that invoice number is not one of mine,
are you sure you brought it to me?
Customer. Well it's got your sticker on the back.
Me. Ah good, what's the number on the sticker?
Customer. xxxxxxx.
Me. (After a quick look up on my haunted fish tank,
computer that is), That refers to a repair 2 years ago,
where I changed a gamma correction chip cause the
picture was all negative looking. Whereabouts did you
take the set recently?
Customer. You guys, the shop on the coast.
Me. Ah. I think you took it to Ponce in Mojacar didn't
you? that's not me. I am nowhere near the coast, as
you can see from the view out the front door.
Customer. It's all the same init? if I bring it to you, you
fix it and get the money from that lot?
Me. Doesn't really work like that does it???
Customer. Well they don't speak English.
Me. (Was going to say, nor do you) Well that's
because they are a Spanish company in Spain and
you have not bothered to learn the lingo. I'm 54 and
probably have not got enough years left to carry on
our conversation. Bring me the set if you want.

Had to write this one up as it shows how the internet
search engines can get it wrong.
Keep in mind my website is quite specific, TVs, HiFi,
electronic repairs etc.....It is really quite simple isn't it?
So, phone call to the office.

Customer. Oh, good morning, don't know if you can help, but I'm having trouble with my turntable.

Me. Ok, what's up with it?

Customer. Well, it's not turning.

Me. Is it part of a large system or stand alone?

Customer. Bloody big.

Me. Ok (thinking it must be an expensive audiophile bit of kit), is it a belt drive or direct, what make is it?

Customer. No idea of the make. Not belt drive, all gear driven.

Me. That's strange. Best pop it in to the office and ill have a look.

Customer. You having a laugh, ill need a f...ing 7 ton truck.

Me. what are we talking about here?

Customer. It's a lorry turntable. Lorries come in, unload and the table turns them round to save driving and reversing in my warehouse.

Me. And you got me details off the net?? I fix record player turntables. Not what you have......

"Good morning, is that the tv fixer bloke at Zeta?"

Yes.

"I have a LG TV that doesn't seem to start when it's cold"

OK, bring it in, probably needs a set of plugs and a distributor cap and rota arm.

"ok, roughly how much would that be?"

Nice start to the day.

Just for light amusement....

Woman came in today to collect her microwave. For what ever reason, she started grumbling about the small ants that she gets in her kitchen. As you do when you go into a shop. Told her to turn the microwave on but open the door for 2 mins. Then the microwaves will spill on the floor and kill the ants. Oh she said, that's a good idea. I asked her if she had a cat, be careful it doesn't burn its paws. No she said, I have a parrot, will that be ok? I said sure, if it's in a cage made by a company called Faraday, it will be good. Ok, she said, I'll check.

Have to say that I get a good share of customer wanting to know what the fault is so they can go off and buy parts openly from sellers such as Flatparts lets say. I make a judgment call on a customer by customer basis. You get to see what they are after, then, I ask for an inspection fee. Often get phone calls where the customer tells you what's wrong and asks if you know what the fault is and how much would it cost. I know these guys are fishing for a remote diagnostic check then they are off to buy the bits. Had one set in and the customer was insistent that I confirmed what was wrong and wanted it back after as he couldn't afford the fix at the moment. I checked the set with a power supply I had to hand and got the set going. Just for amusement, I entered a security pin number in the TV's security settings menu and refitted the old power supply. Gave him the set back as he required. 2 weeks later, another call from someone else with the same model set and serial number, asking if I could talk him through removing the security code. I suggested he bring the set in as it was a very complex job and needed a factory issue remote control. I could hear him grinding his teeth. Made me chuckle.

Is it me?? Really.
I seem to get more than my share of twat customers.
A recent example.
Mr. Total Axxxxxe walks in with a JVC LCD set. "Morning mate, look at my telly, the remote doesn't work". Ok I said, ill book it in. "You have the remote?" "No, I didn't think you would need it". "Ok, well how do you suggest I test and fix it, if I don't have it here?" "Oh, 'spose ill have to go back home and get it". Well I said, It's up to you.
2 hours later, he turns up with the remote. "Here ya go mate". "Thanks" I said. I looked and there were no batteries. I asked the chap if he knew that the batteries were missing. He said "Oh, I took them out". "Any reason?" I said. "Not really" he said. I told him to nip across the road to the supermarket and get a pack of AAAs. "Got me running around a bit avnt ya" says he.

Now, my name is Seth Pittham. I hate both Christian and Sir names. They don't travel over the phone well

and I get called all sorts. A "lady" came into the shop last week. I say lady; she looked like she could kick start 747s. She was clutching the page out of a magazine I write in and advertise in. The advert is quite clear, electronic repairs, TV, audio etc…Really. <u>10% off if you bring the page in with you</u> says the advert. This is how the conversation went. Lady; "Hello, you must be Stiff". Trying to bite my lip I said "Well, it's a bit early in the morning but give me half an hour or so and a cup of coffee and we will see what happens". Lady "er?" "never mind" I said, "my name is Seth, it's a bit difficult" I said. She said "anyway, do you sell car batteries?"….. "Fresh out this morning, had a massive run on them yesterday" was all I could say.
I think it is me being over sensitive.

They walk amongst us.
Mr. Knobby came in a few days ago.
Allo mate, I need one of these…..
This was a 150uF 100v odd capacitor. All swollen and burst.
Having seen this sort of thing many times before, you get to know exactly what the problem is going to be. I said "Ah, I bet that has come from a LG plasma TV, the board on the RHS?
Yeah, that's right, my mate said it's the capacitator and it needs changing.
Ok, I said, it's a SUS board problem and the fault is not with the capacitator as you call it, it's actually a

capacitor, condenser if you are an old fart like me, but with the IPM (big expensive bit that blows up, causing the capacitor to fail). I can sell you the part, but you will need a net with it.

A net? Says bloke.

Yes, I said, when you fit it and turn the set on, it will go pop again after about 10 seconds and the can that it's fitted in will fly off. The net will help you catch it.

No, I bet you want the job, I'll do it.

Fine. 5 € please.

Next morning…….. "That capacitator you sold me went bang. It must have been duff….."

Made me smile this morning.

"Hello, I hear you fix things. My Sky digi-box says technical fault when I try and record".

Ok I said, it sounds like a hard drive fault. Let me have a look and bring the viewing card with you that's in the box, as without that, I can't test the recording mode.

"Ok. Do I need to bring the sky box in to you?"

……… Well if you want me to fix it, it would probably be a good idea. Or you could send me a picture of the faulty box and I'll send you one of a working box.

"Oh, OK, how does that work then?"

Interesting one for a Friday morning.

"Allo mate, my friend says you are a technical mechanic?"

"Well I do try"
Can you have a look at my son's trombone. Be broke the pipe from the bell thing at the end that goes to where the push button are?
"I will see if I can solder it for you...." I then thought... Trombones generally don't have push buttons. Well, some do.

"Good morning, I have an LG plasma TV and in the evenings when it's dark, on dim scenes, I can see little red and green twinkles, stars on the screen".
Now this is a common fault with aging plasma TV sets. Ah, I said, this time of the year when it's clear, you will see the stars coming out. If it's cloudy, you won't see them"
Before I could say any more, the cap said "oh that's fine, I thought it may be fault, bye" and put the phone down. Another happy customer.

You learn something new each day..
Customer "Hello, I have a Samsung plasma or something, and the picture is very very dark and fuzzy. My friend who had a TV shop in the '60s says it needs a new ionizer"
Me "Ummm, I am not sure of a set having an ionizer,

can you give me the model number please? "

Customer "It seems quite simple to me, it just needs an ionizer, and can you get one?"

Me (Why bother I though...) "Yes I said, I have both Mk 1, Mk 2 and a special offer on Mk3s this week, in stock, bring me the TV"

Customer "Oh, I have to bring the set to you?"

Me "Yes, or there is a call out charge"

Customer "It's just getting worse isn't it?"

Me "Not for me" I said.

Ex-Pats thinking in Spain.

Had a call from an Air Conditioning engineer, Colin Baldwin, who I do repairs for and work with. One of the few people who actually knows his trade. "Just given your number to someone who I have just changed an Air Con system for...you'll have fun with this one bud…".

Phone goes and a woman said "Hello, I've got a 32" Grundig TV, and when it's been on for 5 mins, the sound starts to go funny. It's too big to carry, can you come out?".

I went out to be confronted by an old 32" CRT Grundig. The tube as knackered, she watched it with the blinds down as the tube was very low emission. She said "I don't have much money, just spent 500€ on my air con". Now, the reason why she had to replace the air con, was that she had the TV placed under it, to cool the TV down so the sound would be ok! I asked her how long has this been going on...."2 years".

As luck would happen, it was the 18 MHz crystal in the Dolby cct. Warm it with your finger and the sound crackled and went off. Give it a blast of spray cold air from the can of Freezer bought the sound back. I had one and fitted it. All fine.

I sat down with her and on a scrap of paper worked out how much she had spent on electricity running the AC, plus the 500€ recently spent for replacing the compressor in the air con unit. I said you could have bought 3 flat screens by now. "Oh, I didn't want to spend money on the TV.

Oh well......

Customer popped his head in the door just now and said "Those woofer valves you put in my fan are still working"

Sorry I said.

The woofer valves you put in my ceiling fan.

Totally at a loss, after a while I said "Capacitors"

That's right he said.

So there we go. My new name for capacitors now.

How the hell did he make that term up?

This made me laugh today... I generally don't do call-outs for white goods. By the time you make a couple of trips, the bill is too expensive. A woman came into the shop and said "Do you do call outs to fix electric ovens?" I said no, I've done a few in the past, but always got my fingers burnt. "ok" she said and walked out. I thought it was funny.

Sunday evening.... Call from a UK mobile....(I am in Spain, so get the cost of calling back) "Hello, my fridge has stopped working, but the freezer is ok though".(No sorry to bother you on a Sunday evening etc..). After several questions, I asked her what make it was. "I don't know, I'm from Yorkshire". What the hell am I supposed to say to that???

One thing that really annoys me, well one of the many things, customers coming in, asking if they can use my loo.
Generally I say no. However the other day a man and woman came in with a small cheap TV. The lady stood there holding her tummy and asked for the loo. This time I said yes as I was worried that I might have an additional problem to clear up. Later that day, I popped into the loo only to find she had not flushed it, leaving me a couple of pressies.
I did the repair and entered a line on the invoice along with parts and labour, and entry, **ROTSTs** with a 5€ charge, plus VAT of course.
The husband came in and collected his TV. He looked at the invoice, thanked me and asked, "What kind of

component is a ROTST?" I told him it was the Removal of Two Small Turds, as your wife didn't flush the loo after she used it. Oh he said. Paid up, no problems.

Don't you just know that the first job on a Monday morning is likely to set the tone for the week's work?????
Xbox360 comes in. Lady says "The CD won't come out and now it's not reading it"
I booked it in and later that day, moved it from the shelf. As I tipped it forward, drips of something came from underneath. Asking the customer if anyone has done anything, "Oh yeah, my husband sprayed a load of WD40 inside". Bin.

Had a bloke the other morning, asking if I supplied Samsung TV capacitors. Google told him the fault with his TV required these things changing. He had the back off, saw some with bulging tops. Told him I didn't sell parts. He asked "why, you must of had them if you repair sets. They don't cost much, I checked". Fine I said, also check the eeprom as well? What's that? I said ask Google. I said to him if he wanted, I would leave the keys to the workshop under the mat and he can do the set when he wanted. And some people say I am unhelpful???

A couple of things today that made me want to say naughty words the other day.....

Customer calls "Oh hello. Are you qualified to service and repair a Viscount electronic church organ, situated in the Anglican Church in Mojacar?" Let me just add that the tone of the woman's voice was very patronising and Holier than Thou and if nothing else annoying in the extreme. So, I had to reply....."Let me just check my C&G qualifications a minute....
Although I have been in this stupid business for 45 years, servicing everything from bath time toys to GCHQ Troposcater encryption systems, it is important to me that I have the right paperwork.... Ah...I see that I am legally allowed to service this equipment in Spain, but there is an exclusivity clause which states that Juan Pedro Gimenez the 2nd has the rights to repair this very model of Organ, should it be located within 1.2 Km radius of Mojacar. So very sorry, I must revert you to the good Lord above". I heard nothing further.

Phone rings..."Oh good day... Are you in your little workshop this morning?" Patronising bastard I thought. Just having stabbed myself with a screwdriver by accident I was not in the best of moods. I replied.."Indeed I am Sir, if I sound a little quiet, it is because I am talking on a little telephone, doing little jobs for little money. Do you have a little fault or is it big? If it's the latter, I cannot help and suggest you take it to a grown up repair person".

Went quiet for a while…. "Oh have I offended you?"…."let me think of a little reply".

Thursday morning delight.....
"Allo mate, got a LG telly in me car, the starter switch has gone, you know, those white things you put in ya kitchen light. Keeps clicking (Standard LG TV fault), so it's only a small thing". Isn't it nice to have customers that give you as much information and help as possible. Just think of the time I would have wasted, fixing the PSU? I advised him that if he was sure it was the starter switch, nip into the local hardware shop.

Chap comes in..."Morning, I was wondering if you had a chance to look at my Humax box?" Yes, it's fixed, I have tried to call several times, but your mobile just rings out and sent you a text last week......" "Oh" says chappie. " I hate mobiles, I never answer it in case it's a Spanish person, can't do the text thing". "OK" says I. "I'll take your full address and I'll try paranormal remote viewing next time". Chappie says, "Ok, is that an app for the phone?"
How do these people dress themselves in the morning?

Oh we did have a laugh the other day..

Customer in yesterday, one of the nicer ones. Had an older Toshiba LCD that he got 2nd hand. Picture over contrast, so thought ah, that will be a nasty little A15F integrated circuit to replace etc... but no. So got into the engineers menu and started changing some of the settings, eventually got a reasonable picture.

I then saw a setting to turn the image upside down. Some sets have this depending on how you want to mount the set. Now, not having grown up yet, I flipped the setting and called customer to say set was done. I had a contractor in the shop who I had briefed as the customer pulled up. Customer came in and I asked him if he wanted to see the set on. Sure he said. It took him 30 seconds before he said "the pictures upside down". Don't think so, I sad. Contractor looked over and piped up saying what's wrong with that. Managed to keep it going for about 5 mins. Customer by this time was going mad and thought he had a stroke or something. At that point I started laughing and he called me a series of very rude words. The joke bit me on the arse to a degree, as it took me ages to gain access to the engineers menu and to find the setting to put things back to normal. Should have videoed it and popped it on YouTube.
Bet you wish you had as much fun as I do.

"Oh, good morning, could you tell me how I can tell if I have a Smart TV?"
"Sure" I said, "It normally comes with a shirt and tie".

*"Oh, ok, I'll check and call you back if needed.
thanks".*
I thought it was funny for 09.00 Wed morning.

*"aallo mate, told you fix guitar amps. Got a Marshall,
sounds 'orrid when I play my guitar through it".*
Well, I said, tune your guitar and practice more.
*"Funny c..t" was the reply. He put the phone down.
How hurtful.*

Chapter Twenty eight

The Microtron Years

Whilst at the company I served my apprenticeship with, Muirhead, I got to know and became good pals with a chap, Robert Field. He was, still is, a good software and hardware engineer. We had similar interests in radio, he was a licensed Radio Ham. In the late '70s early '80s, CB radio was the in thing. We decided to set up a little business which we called Microtron Electronics. Actually, it was all pucka you know. V.A.T registered etc... We did all sorts. It started by selling and repairing CB radios from the boots of our respective cars at meeting and pub car parks. As things progressed, we developed an idea that Robert had as a result of his relationship with the owners of a well respected Dental Practice in London's West End. This was a system which allowed the reception of the practice to communicate to the Dental Surgeons and visa versa via a system of coded flashing lights. This meant that there was no intercom or telephone activity, putting off the patients. It sounds odd, but it was a great idea. A system was made and installed and worked well for some 20 years. Variations were installed at several other locations.

We made bespoke systems including video and entry systems. At one site we designed a system and installed it. As part of this, various cameras and monitors which were controlled centrally. We fitted a camera to monitor the front door of this particular practice. Now I do not like heights. The camera was fitted high up a wall and cables run. You can imagine the looks on our faces when the system was powered up, only to find the image from the outside camera (in it's weatherproof box) was upside down! Cock. Now being a TV knowledgeable bloke, my suggestion was to bodge, sorry, modify, the TV monitor to show the image inverted. Off with the back off the monitor and I re-soldered the connections to the scan coils of the tube. Great...a picture now the right way up. Robert and I looked at each other in a "job-well-done" manner. But, something I should have considered, this picture was now back to front! Anyone looking at the front door coverage would see the name of the practice and street number in mirror image, reverse. It was cold and now raining. "Leave it" I said. This system stayed in service for some 15 years. Not once did anyone say the image is wrong!

We went on to design alarm systems that would speak to you and all sorts. One venture that was not great was a device we called a "Leak Light". From my musical background I knew that repairing woodwind instruments was a problem. I came up with this idea of a thin fluorescent tube that could be put down the barrel of a saxophone, clarinet, flute etc.... When the keys were pressed, should you see a glimpse of light from the tube, shining from under the key, you knew you had a problem. We sold many to a famous London Woodwind repair company. Unfortunately, I will take responsibility for this one, these were not too reliable and had the rather annoying habit of catching fire. The company we sold to found this feature not to their liking. You just cannot please all the people all the time.

Due to full time work commitments, neither of us had the time to continue with Microtron. That, looking back, was a shame. We could have bought Alan Sugar out at some point. Since then, Robert has grown up and now designs equipment such as lighting desks and control systems.

An advert for the C.B.Radio part of the business we had.
Sketched by my late Brother Jason, who had the knack of taking the you-know-what out of me.

About the author

I guess even before I was born, I knew that I was going to spend a life in the electrical – electronic business. As a child, I loved anything mechanical and electrical.

Growing up in the early 1960s in South London, with parents and Grandparents, I had a good teaching from a Grandfather who was a practical man, making furniture and stuff in the garden shed. A stream of family friends and relatives bought me bits of junk, old radios and TVs to play with. Not an Xbox in sight.

School days were not great. I was not academic and preferred to get on with the practical things. In fact at secondary school in Tulse Hill, Brixton, I managed to get my way into the media Resources Department, where I fixed the school's projectors and duplicating machines. Also repaired teachers radios and TVs. Hardly went to any classes. Three days before my Technical Drawing O Level, I repaired the teacher's old Murphy TV set. During the exam, I got stuck and put my hand up. The teacher came over and scribbled something on my paper. This helped me get an O level in T.D. increasing the number of exams passed by 50%. Result! Apart from the technical stuff at school, the only other interest was music. Some good teachers, the likes of Ron Pite enthused a number of us. This became important in my later life.

Could not wait to get out of school and eventually got an apprenticeship with a company called Muirhead. I was sent to Croydon College to get my C&G in Telecomms and trained well. I ended up working on Facsimile machine design. Stayed there for 10 years. Sent all over the world working on all kinds of equipment connecting facsimile machines to encrypted secure networks, working on HF radio etc.

I left and went to work for Shipton Communications, as a support engineer for their large German made (DeTeWe) PABX telephone system. Good days and ended up as a Director of Engineering. Stayed there for another 10 years.

My daughter Hannah came along in '87. It goes without saying she is the best daughter someone could have. And, unlike me, academic, has a degree in nursing. Brilliant.

In the late '90s, I, along with 3 others set up our own Telecoms and I.T. Company. We did well turning over £3m at one point, but suffered in the down turn and were hit by 9/11. We had a number of service contracts and potential sales with U.S. companies. They all pulled out. We sold the business to another independent I.T company, where I worked for 2 years prior to Coming to Spain.

In the early 80's I bumped into a Dance Band leader, Ken Mackintosh. He was well known in the '50s and continued to run a band. We became friends and he rekindled my earlier love to playing the saxophone. He taught me to play properly and long with a few others, we were allowed to play in his band. We had many years of fun, good times. He sadly passed away shortly before I left for Spain.

The other thing that I forgot to mention was a love for vintage cars. Whilst at the first company I worked for, Muirhead, I became friendly with a Draughtsman. He has a 1951 Rover 75 "Cyclops". I loved the car and eventually bought it from him. It was a sold car and needed some work. The odd thing, to preserve it, he had it painted in Green Hammereite paint. It looked like a poxy green toolbox on wheels. I spent ages rubbing it down and re-spraying it black as it was originally. Loved it.

With my new partner, Amanda, we settled in a backwater village, with a large old townhouse, which continues to slip down the hill where it sits. It soon became apparent that there was a lack of electrical repair services in the area. So, we set up two businesses, Amanda specialising in legal services for the purchase and sale of properties and me fixing TVs and stuff. Throughout my working life, I pursued a passion for the restoration of vintage radio, HiFi and TV.

A friend of mine, Gerry Well had established the Vintage Radio Museum in Dulwich South London during the '50s. A magical place with 1000's of vintage radios TV and all that goes with it. I loved going there and chatting about all this stuff, he was great character. I remember taking my close friend Shane Maxwell there once. Now you have to understand that Shane has not one technical bone in his body, and no interest in radio etc. I introduced Shane to Gerry. Within 5 minutes, Shane fell asleep whilst being shown one of the first combined radiogram-TV sets. Ungrateful bastard.

Sadly there is little restoration work required here, in Spain, but more bread and butter stuff, repairing anything that is brought through the door.

I hope some of the parts in this book were of interest.

Seth

Printed in Great Britain
by Amazon